Lyric Poetry and Space Exploration from Einstein to the Present

Lyric Poetry and Space Exploration from Einstein to the Present

MARGARET GREAVES

OXFORD
UNIVERSITY PRESS

Great Clarendon Street, Oxford, OX2 6DP,
United Kingdom

Oxford University Press is a department of the University of Oxford.
It furthers the University's objective of excellence in research, scholarship,
and education by publishing worldwide. Oxford is a registered trade mark of
Oxford University Press in the UK and in certain other countries

© Margaret Greaves 2023

The moral rights of the author have been asserted

All rights reserved. No part of this publication may be reproduced, stored in
a retrieval system, or transmitted, in any form or by any means, without the
prior permission in writing of Oxford University Press, or as expressly permitted
by law, by licence or under terms agreed with the appropriate reprographics
rights organization. Enquiries concerning reproduction outside the scope of the
above should be sent to the Rights Department, Oxford University Press, at the
address above

You must not circulate this work in any other form
and you must impose this same condition on any acquirer

Published in the United States of America by Oxford University Press
198 Madison Avenue, New York, NY 10016, United States of America

British Library Cataloguing in Publication Data

Data available

Library of Congress Control Number: 2022952222

ISBN 978-0-19-286745-2

DOI: 10.1093/oso/9780192867452.001.0001

Printed and bound by
CPI Group (UK) Ltd, Croydon, CR0 4YY

Links to third party websites are provided by Oxford in good faith and
for information only. Oxford disclaims any responsibility for the materials
contained in any third party website referenced in this work.

for Çağrı

Acknowledgments

Writing this book, with its innumerable and unfathomable little parts, has sometimes felt like trying to fashion a universe. I am grateful to the many people who helped it come together.

For offering material support and intellectual communities, I am grateful to Agnes Scott College, Skidmore College, Hamilton College (especially Christian Goodwillie of the Burke Library), Emory University's Rose Library, New York University, the University of Pittsburgh, Skidmore College's Tang Museum, the New York Public Library, the Yeats Society, the Eliot Society, the Fulbright Commission in Ireland, and the Mellon Foundation. Thank you to my mentors, Deepika Bahri, Christine Cozzens, Ben Reiss, Ronald Schuchard, Nathan Suhr-Sytsma, Rachel Trousdale, and above all Geraldine Higgins, who told me in Sligo to write the book I wanted to write. To my colleagues, especially Cecilia Aldarondo, Lia Ball, Paul Benzon, Anthony Carlton Cooke, Olivia Dunn, Amy Elkins, Catherine Golden, Sarah Harsh, Richie Hofmann, Lucia Hulsether, Nick Junkerman, Michael Marx, Susannah Mintz, Jamie Parra, Jay Rogoff, Molly Slavin, Marlo Starr, Mason Stokes, and Tim Wientzen for their friendship and valuable feedback on parts of the manuscript. And to readers in my more distant orbit for their time and encouragement, especially Stephanie Burt, Cóilín Parsons, Nikki Skillman, and Robert Stilling.

My colleagues and friends Sumita Chakraborty, Bellee Jones-Pierce, Wendy Lee, and Emily Leithauser deserve a special dedication page for the hours and years of conversations, expert interventions, and trust in this project that they provided. I would not have written this book without their humor and brilliance.

I am also grateful for the liberal arts college experiences that ignited this interdisciplinary work. Thank you to the creative faculty of Agnes Scott College, including Chris De Pree, the astronomy professor who encouraged my early efforts at interdisciplinarity. Thank you to Mary Odekon, who supported these ongoing efforts when I joined the faculty at Skidmore. Thank you to my Skidmore students, above all Emilka Jansen, Grace Keir, Olivia Lipkin, and Johanna Meezan for their editorial assistance and enthusiasm. I am also grateful to the diligent and delightful students in my classes Reading the Cosmos (Fall 2018 and Fall 2021), Literature and the Cosmos (Spring

2018), and above all Modernist Poetry (Agnes Scott, Fall 2014)—a group that endured my first experiment of teaching modern poetry in a planetarium.

To the teachers and caregivers who cared for my two young children (through a global pandemic!) while I wrote this book, especially the incredible Jen Marquette, Michelle Boyus, Erica Kemp, and Liza Pennington.

To my family, Kathryn Temple (who is both my aunt and my mentor), Bill Streever, and Kristyn Snedden (my mother, a visionary).

Thank you to my children, Matilda and Silvie, outer space enthusiasts and lights on a troubled planet.

And of course, and most, thank you to Cagri Ozgur, my home in the cosmos.

Contents

List of Illustrations — xi
Permissions — xiii

Introduction: The Lyric and the American Cosmos — 1
   I.1. The Lyric and the Cosmos — 2
   I.2. The Lyric and the Cosmic Scale — 13
   I.3. The American Lyric in Space — 17
   I.4. The Lyric in Cold War Culture — 23
   I.5. The Lyric in the New Space Age — 32

1. The Lyrical Planet: Global Aesthetics and Planetary Ethics — 36
   1.1. The Planet and the Nation — 41
   1.2. Planetary Apostrophe — 45
   1.3. Lyrical Astronauts and Apollo's Eye — 51
   1.4. Lyricizing the Planet — 57
   1.5. Dreaming of Other Worlds — 63

2. "Galaxies of Women": Containment Culture and the Queer Astronomical Lyric — 67
   2.1. The Space Age and the "Breakthrough" Narrative — 69
   2.2. Cold War Surveillance and Queer Opacity — 73
   2.3. Adrienne Rich's Astronomical "We" — 79
   2.4. Elizabeth Bishop's Queer Astronomy — 87

3. "The Moon's Corpse Rising": The Poetic Moon and Imperialist Nostalgia from the U.S. to Kashmir — 101
   3.1. The Lost Poetic Moon — 103
   3.2. Agha Shahid Ali's Lunar Nostalgia — 109
   3.3. The Cold War Ghazal — 118

4. "Out of This World": Cosmopolitanism in Cold War Émigré Poetry — 134
   4.1. Lyric Otherworlds — 136
   4.2. Cold War Émigrés and the Gravity of History — 139
   4.3. Miłosz's Lyric Otherworlds — 148
   4.4. Heaney's Lyric Otherworlds — 152

5. "Vast and Unreadable": Tracy K. Smith, Astronomy, and
   Lyric Opacity in African American Poetry 167
   5.1. Astronomy and African American Poetry 171
   5.2. Lyric Opacity 175
   5.3. Astrophysics, Science Fiction, and Elegy 189

   Coda: Poems in Space 199

Bibliography 203
Index 221

List of Illustrations

1.1. "Earthrise" (1968)	37
1.2. "Blue Marble" (1972)	38
1.3. "Earthrise" (1968) in its original orientation	57

Permissions

"The Country Without a Post Office." Copyright © 1997 by Agha Shahid Ali, "Snow on the Desert." Copyright © 1991 by Agha Shahid Ali, "From Amherst to Kashmir." Copyright © 2002 by Agha Shahid Ali, "The Keeper of the Dead Hotel." Copyright © 1991 by Agha Shahid Ali, "From Another Desert." Copyright © 1991 by Agha Shahid Ali, "Ghazal [In Jerusalem a dead phone's dialed by exiles]." Copyright © 2002 by Agha Shahid Ali, "For You." Copyright © 2003 by Agha Shahid Ali Literary Trust, from THE VEILED SUITE: THE COLLECTED POEMS by Agha Shahid Ali. Used by permission of W. W. Norton & Company, Inc.

"Postcard from Kashmir," from *The Half-Inch Himalayas* (Middletown, CT: Wesleyan University Press, 1987), © 1987 by Agha Shahid Ali Literary Trust. Used by permission of Wesleyan University Press.

"Lunarscape," from *Modern Indian Poetry in English*, ed. P. Lal (Kolkata Writers Workshop). Copyright © 1971 by Agha Shahid Ali Literary Trust. Excerpts from *In Memory of Begum Akhtar*. Copyright © 1979 by Agha Shahid Ali Literary Trust.

"We Who Are Executed," from *The Rebel's Silhouette: Selected Poems*, by Agha Shahid Ali and Faiz Ahmad Faiz (Amherst, MA: University of Massachusetts Press, 1995). Used by permission of University of Massachusetts Press.

"A Brave and Startling Truth" from CELEBRATIONS: RITUALS OF PEACE AND PRAYER by Maya Angelou, copyright © 2006 by Maya Angelou. Used by permission of Random House, an imprint and division of Penguin Random House LLC. All rights reserved.

Maya Angelou, "A Brave and Startling Truth," in *Maya Angelou: The Complete Poetry*, Little Brown Book Group Ltd., copyright © 2006 by Maya Angelo. Reproduced with permission of the Lincensor through PLSclear.

Excerpts from W. H. Auden, *The Collected Poems*. Copyright © 1940, 1958, 1968, 1967, 1969 by W. H. Auden, renewed. Reprinted by permission of Curtis Brown, Ltd. All rights reserved.

"The More Loving One," copyright © 1957 by W. H. Auden and renewed 1985 by The Estate of W. H. Auden; and "Ode to Terminus," copyright © 1968 by W. H. Auden; from COLLECTED POEMS by W. H. Auden, edited by

Edward Mendelson. Used by permission of Random House, an imprint and division of Penguin Random House LLC. All rights reserved.

"The Armadillo," "In the Waiting Room," "Insomnia," "The Shampoo" from *Poems*. Copyright © 2011 by Alice H. Methfessel Trust. Used by permission of Farrar, Straus and Giroux, LLC.

"Geometry" from *The Yellow House on the Corner*, Carnegie Mellon University Press, Pittsburgh, PA. © 1980 by Rita Dove. Used by permission of the author.

Fatimah Asghar, "Pluto Shits on the Universe," from *Poetry*. © 2015 by Fatimah Asghar. Used by permission of the author.

From "Little Gidding" and "Burnt Norton," from *The Complete Poems and Plays: 1909–1950* by T. S. Eliot. Copyright 1950, 1943, 1939, 1930 by T. S. Eliot. Copyright 1952, 1936, 1935 by Harcourt Brace & Company. Copyright renewed 1964, 1963, 1958 by T. S. Eliot. Copyright renewed 1980, 1978, 1971, 1967 by Esme Valerie Eliot. Used by permission of HarperCollins Publishers and Faber & Faber, Ltd.

"[American Journal]." Copyright © 1978, 1982 by Robert Hayden, from COLLECTED POEMS OF ROBERT HAYDEN by Robert Hayden, edited by Frederick Glaysher. Used by permission of Liveright Publishing Corporation.

"The Flight Path" from *The Spirit Level*. Copyright © 1996 by the Estate of Seamus Heaney. "Alphabets," "The Birthplace," "*Clearances* VIII," "The Cure at Troy," "Digging," "Kinship," "The Tollund Man," "Whatever You Say, Say Nothing," and "Westering" from *Opened Ground: Selected Poems, 1966–1996*. Copyright © 1998 by the Estate of Seamus Heaney. "Out of This World" from *District and Circle*. Copyright © 2006 by the Estate of Seamus Heaney. Used by permission of Farrar, Straus and Giroux, LLC, and Faber & Faber Ltd.

From "Voyage to the Moon" from *Collected Poems 1917 To 1982* by Archibald MacLeish. Copyright ©1985 by The Estate of Archibald MacLeish. Used by permission of HarperCollins Publishers.

"Overdue Pilgrimage to Nova Scotia" from COLLECTED POEMS by James Merrill, copyright © 2001 by the Literary Estate of James Merrill at Washington University. Used by permission of Alfred A. Knopf, an imprint of the Knopf Doubleday Publishing Group, a division of Penguin Random House LLC. All rights reserved.

From "Campo Dei Fiori," "Incantation," "ArsPoetica," "An Appeal," from New and Collected Poems 1931–2001 by Czeslaw Milosz. Copyright © 1988, 1991, 1995, 2001 by Czeslaw Milosz Royalties, Inc. Used by permission of HarperCollins Publishers.

From "A Birthday Present" and "Wuthering Heights," from *The Collected Poems*, Sylvia Plath. Copyright © 1960, 1965, 1971, 1981 by the Estate of Sylvia Plath. Editorial material copyright © 1981 by Ted Hughes. Used by permission of HarperCollins Publishers and Faber & Faber Ltd.

The lines from "The Explorers." Copyright © 2016 by the Adrienne Rich Literary Trust. Copyright © 1955 by Adrienne Rich, The lines from "Planetarium." Copyright © 2016 by the Adrienne Rich Literary Trust. Copyright © 1971 by W. W. Norton & Company, Inc, from *COLLECTED POEMS*: 1950–2012 by Adrienne Rich. Used by permission of W. W. Norton & Company, Inc.

Tracy K. Smith, "My God, It's Full of Stars," "The Universe Is a House Party," "Don't You Wonder, Sometimes," and "Sci-Fi" from Such Color: New and Selected Poems. Originally collected in *Life on Mars*. Copyright © 2011 by Tracy K. Smith. Reprinted with the permission of The Permissions Company, LLC on behalf of Graywolf Press, Minneapolis, Minnesota, www.graywolfpress.org.

From *Captain Cook in the Underworld* by Robert Sullivan (Auckland University Press, 2003). Used by permission of Auckland University Press.

Parts of Chapter 1 and Chapter 3 appeared as "Nostalgic Forms: Agha Shahid Ali, Space Age Poetry, and the Cold War Planet" in *College Literature* 48:3 (Summer 2021): 341–73. Reproduced with permission.

Part of Chapter 5 appeared as "'Vast and Unreadable': Tracy K. Smith, Astronomy, and Lyric Opacity in Contemporary Poetry" in *Contemporary Literature* 61:1 (Spring 2020): 1–31. Reproduced with permission.

Introduction

The Lyric and the American Cosmos

1,100 haiku are currently orbiting Mars. In 2013, NASA's "Going to Mars" educational outreach campaign held a competition to select haiku to fly aboard the Mars Atmosphere and Volatile EvolutioN (MAVEN) craft.[1] NASA received 12,530 submissions from around the world and flew all haiku that got at least two votes, framing the contest as a process of global democracy. The haiku was the perfect choice, formally and culturally. Its brevity and deceptive simplicity make it both portable and user-friendly: more haiku could fly than, say, sonnets, and *anyone*, as many U.S. schoolchildren are taught, can write a three-line poem by counting out syllables. The five haiku that received the most votes in this "global" competition were from the U.S., the UK, and Europe, following the routes along which Ezra Pound popularized the form in the early twentieth century. The competition, in fact, uses the Japanese haiku to present the U.S. space program as an endeavor on behalf of all humankind. The winner of the Mars haiku contest, by British poet Benedict Smith, expresses less optimism about a human collective: "It's funny, they named / Mars after the god of war / Have a look at Earth."[2]

This book brings together lyric studies and the poetry and poetics of the post-World War II era, a period in which the U.S. has styled itself as the universe through cultural narratives about space science. The role of space exploration in American nationalism is obvious enough. "We are masters of the universe," as one senator put it at the dawn of the Cold War space race, emphasizing American values of freedom and choice: "We can go where we choose."[3] While nationalism frequently draws on the cosmic, from Louis XIV (the "Sun King") to the United States Space Force, lyric poetry and astronomy

---

[1] Sarah Loff, "Haikus Selected To Be Carried On NASA's Next Mission to Mars," 8 Aug. 2013, https://www.nasa.gov/content/haikus-selected-to-be-carried-on-nasas-next-mission-to-mars/.
[2] Jennifer Lai, "We're Bringing 1,100 Haiku to Mars This Fall," *Slate*, 9 Aug. 2013, https://slate.com/technology/2013/08/maven-haiku-nasa-s-spacecraft-will-be-bringing-1100-poems-to-mars-this-november.html.
[3] Quoted in Matthew D. Tribbe, *No Requiem for the Space Age: The Apollo Moon Landings and American Culture* (Oxford: Oxford University Press, 2014), 3.

might seem to have little to do with one another. Yet influential critical stories about the lyric—stories that have shaped contemporary American poetry and poetics—emerge from the same tangled cosmologies that sent the U.S. into extraterrestrial space. Excavating lyric poetry's surprising political life as the U.S. has extended its power across and even beyond the globe, *Lyric Poetry and Space Exploration from Einstein to the Present* works at the intersection of poetry, astronomy, and politics to re-evaluate the postwar lyric.

Anchored in the postwar U.S. but informed by a long history of poetry and astronomy, this book argues that contemporary poets look to extraterrestrial space to grapple not only with national politics but also with the politics of the lyric itself. I explore how a transnational range of poets use astronomical language and methods to develop their poetics in the context of Anglo-American imperialism. Poets including Elizabeth Bishop, Adrienne Rich, Seamus Heaney, Derek Walcott, Agha Shahid Ali, and Tracy K. Smith turn to the extraterrestrial to address a set of formal and ethical concerns as they write within and against discourses of American and lyric exceptionalism. This book considers their work alongside a range of sources that pair astronomy with lyric poetry, including popular magazines, NASA materials, space photography, and television. This archive reveals the lyric's complex cultural roles in at once extending and undermining expansionist fantasies. Dazzled by the aesthetics of astronomy but wary of its imperial uses, poets invoke the strangeness of astronomical figures to envision more ethical modes of subjective experience and collective belonging.

I.1 The Lyric and the Cosmos

At the origin of the Anglo-American empire is a cosmic metaphor. In 1768 Captain Cook sailed for the South Pacific aboard the *Endeavour* to track the transit of Venus across the Sun. The voyage of the *Endeavour* aimed to map the universe in a European image—which is to say, it universalized Europe by seizing indigenous lands in the name of bettering humankind through scientific progress. In *Transit of Empire*, Jodi Byrd follows the movements of Anglo-American empire through the astronomical figure of transit beginning with Cook's celestial voyages. Her term "imperialist planetarity" describes the alignment of scientific discourse, geographical mastery, and universality that led Cook's voyage to write Eurocentric values into the design of the night sky.[4]

---

[4] Jodi A. Byrd, *The Transit of Empire: Indigenous Critiques of Colonialism* (Minneapolis: University of Minnesota Press, 2011), xx.

That tracking Venus "was the purpose, or cover, for what followed Cook reveals...something telling about the nature of British and American colonialism and imperialism that remained allied even during the family squabble that was the American Revolution," she writes.[5] The American Revolution itself relies on a celestial idiom; as Eran Shalev notes, the cosmic term "revolution" made the rise of the American nation seem as natural as the motions of planetary bodies.[6] The shape of an astronomical revolution, moreover, completes a circuit: following a departure from an origin point, there is a full return, describing the move from the U.S. as a colony to an imperial power. The Continental Congress Flag Resolution of 14 June 1777, inscribed the U.S. into the natural laws of the universe through the decree "That the Union be Thirteen Stars, white in a Blue Field, representing a new Constellation." Through imperialist planetarity, the Anglo-American empire has styled itself as the natural ruler of the planet, from the voyage of the *Endeavour* to the movement of manifest destiny into the "final frontier" of space. NASA named a shuttle the *Endeavour*, and the opening monologue of *Star Trek* (which begins "Space: the final frontier") borrows language from Cook's journals, indicating how deeply Cook's cosmic-imperial voyages infuse postwar space culture.[7]

The imperial voyage of the *Endeavour* followed the aesthetic logic of *cosmos*, the ancient Greek word for both "universe" and "order," out of which ideas of poetic design (and particularly the lyric) emerge. Indeed, the first astronomer to record the transit of Venus, Jeremiah Horrocks in 1639, did so in a long poem, *Venus in sole visa*, aligning cosmic harmony with poetry.[8] The ancient Greek philosophical concept of the harmony of the spheres underlies this association. The harmony of the spheres maintained that celestial bodies produce music through their motions, an idea derived from the Pythagoreans' discovery of the mathematically calculable relationship between musical pitch and the length of a string. The harmony of the spheres is frequently symbolized by the poet-prophet Orpheus, who mastered the lyre to produce the song of the universe, or *musica universalis*, which corresponded to the proper

---

[5] Byrd, *The Transit of Empire*, 2.
[6] Shalev writes, "Revolution itself was, of course, an astronomical concept originally derived from the cycles of stars" (48). His essay offers an excellent overview of American nationalism and celestial imagery in the early republic. Eran Shalev, "'A Republic Amidst the Stars': Political Astronomy and the Intellectual Origins of the Stars and Stripes," *Journal of the Early Republic* 31 (Spring 2011): 39–73.
[7] Captain James Cook reflected that his "ambition leads me not only farther than any other man has been before me, but as far as I think it possible for man to go," echoed in the opening monologue's pledge "to boldly go where no man has gone before." James R. Cook, *The Journals of Captain Cook*, ed. Philip Edwards (New York: Penguin, 2000), 331.
[8] See R. Horrocks, "Jeremiah Horrocks, astronomer and poet," *Journal of the Royal Society of New Zealand* 42:2 (2012): 113–20.

tuning of the soul.[9] In his 2003 libretto *Captain Cook in the Underworld,* Maori Robert Sullivan directly links the Western Orphic lyric to the *Endeavour*'s voyage:

> bright Orpheus of the singing lyre,
>   poet exemplar, inspired
> by His Majesty's bark *Endeavour:* a choir
> for *Endeavour* travelling to the transit of Venus.[10]

Sullivan invokes the harmony of the spheres as he imagines the Orphic lyric and the imperial ship traveling in sync to the motions of Venus. Responding to the fusion of Western cosmology and the Western lyric, Sullivan writes poetry out of an entangled cosmology in which the human moves *with* the cosmos, in contrast to the Western ideal of the fixed lyric "I" that interprets the universe revolving around it.[11]

In the second century CE, Ptolemy consolidated Greek cosmology into a geocentric figure of the universe that endured through the scientific revolution and continues to inflect lyric poetry. In the Ptolemaic model, the mutable, material Earth rests at the center of the universe surrounded by rotating celestial spheres. While the Earth is made of changeable substance, the perfect celestial spheres are made of an unchanging, eternal substance called ether. Christian theologians translated this figure of the universe into the perfection of God's mind and creation.[12] As Dante describes in his cosmological epic *The Divine Comedy,* poetic design records how "Among themselves all things / Have order; and from hence the form, which makes / The universe resemble God."[13] For Dante, the poet should attend to "the figuring of Paradise" rather

---

[9] For the Orpheus legend's relationship to the music of the spheres and its impact on subsequent poetry, see Elisabeth Henry, *Orpheus and His Lute: Poetry and the Renewal of Life* (Carbondale and Edwardsville: Southern Illinois University Press, 1992), especially 83–101.

[10] Robert Sullivan, *Captain Cook in the Underworld* (Auckland, NZ: Auckland University Press, 2003), 3.

[11] For an excellent study of astronomy and cosmology in Sullivan's work, see Chadwick Allen, *Trans-Indigenous: Methodologies for Global Native Literary Studies* (Minneapolis: University of Minnesota Press, 2012), especially 193–248.

[12] Multiple cogent histories of astronomy document these ideas. See, for instance, John North's *Norton History of Astronomy and Cosmology* (New York: Norton, 1995), and Norriss S. Hetherington, ed., *Cosmology: Historical, Literary, Philosophical, Religious, and Scientific Perspectives* (New York: Garland, 1993), especially his introduction to the Greeks' geometrical cosmos, 69–77. Astronomer Carl Sagan's chapter "The Harmony of Worlds," in *Cosmos* (New York: Ballantine, 1980), based on his television series of the same name, offers a succinct overview of this cultural history from the Pythagoreans through Newton (41–75).

[13] Dante Alighieri, *Paradiso* I, 103–5, trans. Henry Francis Cary (Roseville, CA: Dry Bones Press, 2000).

than the politically tumultuous, broken material world of fourteenth-century Europe.[14] Three centuries later, John Milton likewise seized on the relationship of poetry to cosmic order in a period of political turmoil. In 1638, Milton visited a Galileo under house arrest, confined by the Inquisition for his support of a heliocentric cosmos. Galileo references abound in *Paradise Lost*, written as Galileo's telescopic observations overturned the Aristotelian–Ptolemaic figure of a geocentric universe framed by immutable cosmic spheres.[15] The epic marches forward in iambic pentameter to invoke divine cosmic order. Wondering at "Their Starry dance in numbers that compute," Milton surveys the heavens in blank verse that computes as perfectly as the Pythagorean planetary-musical intervals described by the harmony of the spheres.[16]

In Cold War America, Vladimir Nabokov made the same cosmic-metrical pun while in political exile. In *Pale Fire* (1961), the poet John Shade (perhaps the alias of Eastern European exiled king Charles Kinbote) aligns inner and outer harmony in his 999-line poem in heroic couplets: "And if my private universe scans right, / So does the verse of galaxies divine. / Which I suspect is an iambic line."[17] From Dante to the fictional Kinbote, poets invoke cosmic harmony to explore the relationships between inner and outer experience as well as to imagine a chaotic world mended into an orderly whole. This poetic-cosmic legacy infuses Walt Whitman's astronomy poems of the Civil War, astronomer Carl Sagan's lyrical popularization of the cosmos during the Cold War, the invocation of Romantic poetry in Gene Roddenberry's *Star Trek*, and NASA's frequent use of lyric to present space exploration as a peaceful mission serving humanity. From Kinbote to Captain Kirk to Neil Armstrong, poetry is bizarrely bound to postwar politics.

Especially in the context of space exploration, the lyric often works to universalize American culture and to sponsor imperialism. As it travels through civic and cosmic space, the lyric functions as one of the "literary, cultural, and aesthetic *genres*" Lisa Lowe identifies "throughout which liberal notions of person, civic community, and national society are established and upheld."[18] The lyric, with its deep investment in questions of the individual and the nation, signals the values of democracy, hence its presence at presidential

---

[14] Dante, *Paradiso* XXIII, 61.
[15] For a definitive study of Milton and cosmology, see Dennis Danielson, *Paradise Lost and the Cosmological Revolution* (Cambridge: Cambridge University Press, 2014). For Milton's relationship with Galileo (the only contemporary he mentions in *Paradise Lost*), see especially 78–128.
[16] John Milton, *The Complete Poems* (New York: Penguin, 1999), 185.
[17] Vladimir Nabokov, *Pale Fire* (New York: Vintage, 1962), 69.
[18] Lisa Lowe, *The Intimacies of Four Continents* (Durham, NC: Duke University Press, 2015), 4.

inaugurations, public events of national mourning, and shuttle launches. Yet little scholarship has addressed the role of the lyric in upholding American nationalism. The sole monograph on lyric form and American imperialism, Jen Hedler Phillis's *Poems of the American Empire*, details how poets manipulate lyric and narrative temporal structures to respond to experiences of living through "their nation's imperial adventure."[19] Phillis's approach, however, differs from mine; she concentrates on the lyric-narrative dialectic in specific poems, rather than analyzing how the *idea* of the lyric operates in civic contexts. I demonstrate how the lyric's ceremonial dimensions and formal features make it an ideal genre through which to sanctify and universalize national and imperial events, from inaugurations to moon landings. But in tracing the relationship between the lyric and space exploration in the Cold War and after, this book also distinguishes between an entrenched cultural imaginary of lyric poetry and the often very different work specific poems do, even as these categories are permeable. The poets explored here often grapple with the politics of the lyric through writing cosmic poetry. Working both within and against the cultural reception of their genre, poets contend with how to experience both "I" and "we" and the size and significance of the "I" in relation to the universal—even the extent to which it is possible or desirable to imagine a universal humanity—through astronomical and cosmological tropes.

Throughout this book, I use the terms "the lyric," "the lyric mode," and "lyricization" or "lyric reading" to describe distinct (though often overlapping) phenomena. Many non-lyric poems and other cultural forms have lyrical properties and do the cultural work of the lyric genre without being lyric poems. For instance, Chapter 1 argues that early photographs of planet Earth taken from space—which the mainstream media often published alongside lines of lyric poetry—are lyrical objects. Their inclusion with lines of poetry in newspapers and magazines "lyricized" them, which is to say, prompted the public to read them as lyric poems. But these photographs also have aesthetic properties associated with apostrophe and lyric temporality that encouraged this association in the first place. While it would certainly be possible to analyze the relationship between poetry broadly construed and astronomy, it is the particular ways that the lyric (as both a genre and a mode) gets shaped, imagined, and valued—including its perhaps historically exaggerated ties to

---

[19] Jen Hedler Phillis, *Poems of the American Empire: Lyric Form in the Long Twentieth Century* (Iowa City: University of Iowa Press, 2019), 6.

ancient Greek aesthetics—that make it such a compelling frame through which to consider postwar Anglo-American imperialism.

Characterizing the lyric as a genre is a contested critical move in lyric studies, but in doing so I mean to excavate rather than occlude key historical and political phenomena. Historical and materialist studies of the lyric, broadly constellated as the New Lyric Studies, have debunked the idea that the lyric is a distinct genre defined by the personal utterance of private emotional experience. The initiators of the New Lyric Studies, Virginia Jackson and Yopie Prins, have demonstrated how "the lyric" is an anachronistic invention of the Romantics, distilled by ahistorical twentieth-century reading practices such as the New Criticism. Through acts of reading, all poetry across its distinct forms and genres became "lyricized," which is to say read as specifically *lyric* poetry, flattening a range of distinct forms of relatively short poetry into "the genre of personal expression" to be read as the "utterance overheard" of a solitary lyric "I."[20] From this perspective, the lyric is a historically irresponsible designation at best, an endorsement of Western universalism and the bourgeois transcendent subject at worst. Yet there are many reasons to keep the concept of the lyric in play even in a historical study like this one. For one, it *is* possible to identify properties that link subtypes of many relatively short poetic forms that this book engages, including the ghazal, the elegy, and the sonnet; identifying these patterns allows for productive comparisons across these distinct forms. Moreover, these lyrical properties add up to a more capacious understanding of the lyric than that identified by Jackson and Prins. In his redress of the New Lyric Studies, for instance, Jonathan Culler characterizes lyric poems as "short, nonnarrative, highly rhythmical productions, often stanzaic, whose aural dimension is crucial." To this list of qualities we could reasonably add an implied or stated "I-thou" address (for Culler, apostrophe is *the* preeminent figure of the lyric), temporal nonlinearity, and an emphasis on formal compression. This loose definition of the lyric does not insist on fixed essences; rather, as Culler puts it, a genre is "a historically evolving set of possibilities with potential to surprise."[21] Indeed, rather than obscuring historical processes, genre brings into focus the shifting, complex relationships between form, value, and culture. Keeping genre in play generates surprising questions at the intersection of the lyric and the culture of

---

[20] Virginia Jackson and Yopie Prins, eds., "Introduction," in *The Lyric Theory Reader: A Critical Anthology* (Baltimore, MD: Johns Hopkins University Press, 2014), 2–3. For a discussion of the historical collapsing of genres that produced the "lyricization" of poetry, see Virginia Jackson, *Dickinson's Misery: A Theory of Lyric Reading* (Princeton, NJ: Princeton University Press, 2005), 1–15.

[21] Jonathan Culler, *Theory of the Lyric* (Cambridge, MA: Harvard University Press, 2017), 89.

space exploration. For instance, why did 1960s space culture revive the Romantic lyric poem, why did Carl Sagan think the fusion of scientific and lyric values could prevent nuclear disaster in Cold War America, and why does Tracy K. Smith interweave lyrical and science fictional elements to elegize her father, who worked on the Hubble Space Telescope? These are questions for genre studies, as they register the particularities of the cultures the lyric has emerged out of and commented on since the 1960s.

This book also keeps genre in focus due to the peculiar entanglement of the lyric and cosmological inquiry—an association that has led to a pervasive sense that the lyric somehow transcends, or attempts to transcend, the material world. Classic and disputed claims about key features of the lyric, particularly its association with immaterial realms, emerge from Greek cosmology, which inflects the cultural imagination of the lyric in ways that have not been adequately understood. The Western lyric abounds with metaphors of the eternity of poetic creation, locating poetic utterance in a crystalline sphere away from Earth's mutability, where not even the longest-lasting forms of the material Earth will outlast poetry's timelessness. (As Shakespeare put it, "Not marble, nor the gilded monuments / Of princes shall outlive this powerful rhyme.")

In fact, three of the most influential descriptions of the lyric—the New Critical metaphor of the "well-wrought urn," Bakhtin's theory of monologic voicing, and Sharon Cameron's study of lyric time—depend on cosmological models. Many of the New Critics specifically invoked the Greeks to describe lyric immateriality and eternity. For instance, Cleanth Brooks's *The Well Wrought Urn* styles itself after Keats's "Ode on a Grecian Urn," which presents a lone speaker addressing an object of both art and death that becomes a metaphor for the self-contained and eternal poem. Brooks developed and extended a method of formalist poetic analysis that treats the lyric poem as a closed, perfectly harmonized, eternal world of its own, belonging to an ethereal cosmic sphere.[22] Bakhtin made the lyric-cosmic association even more clearly in his influential account of genre that privileges the novel. In the decade before the Cold War, he famously characterized poetry as "monologic," distinguishing it from the preferred heteroglossic form of the novel that he found to be more equipped to contend with the competing discourses of modern life. Bakhtin identified the solipsistic utterance of a single voice

---

[22] Cleanth Brooks, *The Well Wrought Urn: Studies in the Structure of Poetry* (New York: Harcourt Brace, 1947). Brooks describes Keats's urn in terms of eternity, "for, like eternity, its history is beyond time, outside time, and for this very reason bewilders our time-ridden minds" (164). He goes on to discuss W. B. Yeats's inheritance of this idea of the eternal poem through Platonic ideal forms (184).

with an outmoded pre-Copernican cosmology; the "special unitary and singular language of poetry," he argued, is "a Ptolemaic conception of the linguistic and stylistic world."[23] Lyric poetry, he implies, is as obsolete as a geocentric figure of the cosmos and, more damningly, just as solipsistic: the monologic poem suggests that the universe revolves around a single speaker. Sharon Cameron echoes Bakhtin's account of pre-Copernican lyric solipsism in her influential *Lyric Time*, where she charts the lyric's rhetorical claims to eternity. Cameron argues, "The lyric is seen as immortal... because it is complete/completed in and of itself, transcending mortal/temporal limits in the very structure of its articulation." The lyric lends its singular speaker "the illusion of alone holding sway over the universe, there being, for all practical purposes, no one else, nothing else, to inhabit it."[24] Echoing Bakhtin's monologic language of poetry, Cameron's account of lyric time positions a singular speaker at the illusory center of a sealed-off poetic universe that presents itself as immune to the decay of the material world. The singular lyric speaker commanding the cosmos is a near parody of the apolitical postwar lyric that, as Clare Cavanagh has wryly put it, "must be coerced back into society by all necessary means."[25] The political "sphere" itself, from which the lyric is often imagined to be divorced, is a Greek cosmological metaphor drawing from distinct crystalline spheres.

These cosmic formulations of the lyric all seem ready-made for one of the best-known cosmological poems, John Donne's aubade "The Sun Rising." Donne's lyric "I," dismayed by the "new philosophy" of heliocentrism that "calls all in doubt," commands the Sun back to its proper sphere to shine on the lovers at their rightful place in the center of the universe: "Shine here to us, and thou art everywhere; / This bed thy center is, these walls, thy sphere."[26] The lovers retreat from the public, political sphere to become the center of the cosmos: "She's all states, and all princes, I, / Nothing else is." These lines, particularly in the context of Britain's expanding empire (evoked in a reference to the spice trade), exemplify the common twentieth-century

---

[23] M. M. Bakhtin, *The Dialogic Imagination: Four Essays* (Austin: University of Texas Press, 2017), 288.
[24] Sharon Cameron, *Lyric Time* (Baltimore, MD: Johns Hopkins University Press, 1979), 119.
[25] Clare Cavanagh, *Lyric Poetry and Modern Politics: Russia, Poland, and the West* (New Haven, CT: Yale University Press, 2010), 10.
[26] John Donne, *An Anatomy of the World*, in *John Donne: The Complete English Poems*, ed. A. J. Smith (London: Penguin, 1996), 276; and Donne, "The Sun Rising," 80. Donne published *An Anatomy of the World* in 1611, the year after Galileo's *Starry Messenger* described Jupiter's moons and the stars of the Milky Way observable only by telescope. The poem, like the telescopic observations described in *Starry Messenger*, is a study in scale and subjectivity. The "new philosophy" refers at once to the new astronomy and to the Reformation that devastated England and Donne's personal life. "The Sun Rising" (1633) has a more playful relationship to the "new philosophy."

critique of bourgeois lyric aestheticism, when the privileged lyric "I" asserts its universality.[27] The cosmic idiom is built into many lyric poems themselves and into modern theories of the lyric, many of which critique its privileged claims to vault individual experience to universal human truths. Of course, "The Sun Rising" hinges on the fact that the speaker cannot actually command the cosmos and has only apostrophe, employed with poignant humor, to capture and elongate a single moment in bed with a lover against a vast, changing universe that has almost eclipsed the "I." In doing so, the poem seems to present a standoff between the sciences and poetry. But here, too, the situation is more complicated: Donne has digested recent cosmological developments and takes them seriously, recording them in the form of the poem, from the relative cosmic positioning of the lovers and the Sun to the poem's play with the language of measurement, scalar contrasts, and visual observation. This cosmological lyric poem at once embraces and subverts the alignment between cosmic and poetic design, as well as the tensions between scientific and lyrical knowledge.

The ostensible divide between science and lyric poetry is itself a question of genre. Jackson and Prins (drawing from Gérard Genette) locate the invention of the lyric as the genre of an emotive individual speaker in the nineteenth-century moment in which the Romantics inaccurately took the lyric to be one of Aristotle's three natural genres.[28] This is also the moment in which poetry was ostensibly transformed into *lyric* poetry through a divorce from the sciences. Keats's "Do not all charms fly / At the mere touch of cold philosophy?" seems to evince this split.[29] Keats's association of "charms" with poetic language and coldness with epistemological inquiry accords with his concept of negative capability, by which he means poetry's ability to dwell "in uncertainties, mysteries, doubts, without any irritable reaching after fact and reason."[30] This apparent division between the epistemological certainties of "cold philosophy" and the mysteries of lyrical domains has organized the cultural imagination of the lyric. Yet, as Amanda Goldstein argues in *Sweet Science*,

---

[27] The lyric's unconscious association with Greek cosmology has also been a source of critical mistrust; in the postwar era, the lyric, with its ostensible focus on individual subjectivity and rhetorical claims to transcend the material world, is often taken to be the worst offender of what Terry Eagleton describes as "the familiar bogeyman of universality," part of "a Western conspiracy which speciously projects our local values and beliefs on to the entire globe." Terry Eagleton, *After Theory* (New York: Basic Books, 2003), 160.

[28] Gérard Genette, "The Architext" (1979; trans. 1992), in Jackson and Prins, eds., *The Lyric Theory Reader* (Baltimore, MD: Johns Hopkins University Press), 17–29.

[29] John Keats, "Lamia," in *John Keats: The Major Works*, ed. Elizabeth Cook (Oxford: Oxford University Press, 1990), 320.

[30] Keats described negative capability in a letter to his brothers, George and Thomas Keats, in December 1817. *John Keats: The Major Works*, 370.

understanding negative capability as a lyrical-scientific standoff misrepresents the Romantic lyric and, indeed, complex Romantic understandings of scientific inquiry. Goldstein develops a theory of materialist figuration at precisely the historical moment that assigned "poetry its enduring place at the prestigious margins of cultural life—as a discourse expert in the immaterial reaches of subjectivity and the special materiality of language itself." Against "the now familiar division of labor between literary and scientific representation" that occurred at this moment, Goldstein "uncovers a countervailing epistemology that casts poetry as a privileged technique of empirical inquiry."[31] By emphasizing lyric figures, Goldstein demonstrates how lyric inquiry became a method for apprehending an unfathomable cosmos. The tasks of lyric poetry and the sciences, the lyric and astronomy, are often surprisingly similar.

Lyric form and rhetoric often emphasize non-normative, unusual, and unassimilable combinations, effects intensified when lyric comes into proximity with astronomy. This book critiques the ways in which government institutions have used the lyric to further political agendas, even as it emphasizes the astronomical lyric's strangeness and disruptive potential. This countervailing tradition of estrangement in the lyric unsettles imperialist claims to universality. As Omaar Hena writes in *Global Anglophone Poetry*, "Poetry's capacity to unsettle and estrange needs emphasizing" in the context of "the enduring inequalities and racial exclusions held over from colonial history, exclusions that work in tandem with liberal, capitalist discourses of development and progress."[32] Avant-garde poetry is more associated with this sort of disruption than the lyric, drawing as it does on Viktor Shklovsky's sense of poetry's "roughened, impeded language" related to *ostranenie* (or Brechtian estrangement).[33] In its association of estrangement with difficulty, avant-garde poetics deliberately foregrounds linguistic alterity. Jacob Edmond, for instance, has demonstrated how cross-cultural avant-garde poetic encounters destabilize the neat oppositions of the Cold War globe.[34] But the avant-garde is not the focus of this book: I am concerned instead with poems that foreground the strangeness of traditional lyric tools, in particular lyric temporality and modes of address. Rather than concentrating on poets who reject lyricism, then, I emphasize those who explore and resist the lyric's assimilative

---

[31] Amanda Jo Goldstein, *Sweet Science: Romantic Materialism and the New Logics of Life* (Chicago, IL: University of Chicago Press, 2017), 7.

[32] Omaar Hena, *Global Anglophone Poetry: Literary Form and Social Critique in Walcott, Muldoon, de Kok, and Nagra* (New York: Palgrave Macmillan, 2015), 5.

[33] Viktor Shklovsky, "Art as Technique," in Julie Rockin and Michael Ryan, eds., *Literary Theory: An Anthology* (Oxford: Blackwell, 2004), 15–21, here 16.

[34] Jacob Edmond, *A Common Strangeness: Contemporary Poetry, Cross-Cultural Encounter, Comparative Literature* (New York: Fordham University Press, 2012).

functions from within its own forms and tools. They work uneasily within the contradictions of lyric form and politics laid bare by the extraterrestrial. My archive investigates how poets from identity groups dehumanized in the making of a harmonious American experience and traditionally elided from or tokenized in astrophysics—including queer poets and poets of color, as well as poets from postcolonial or immigrant positions who critique American exceptionalism—draw on astronomy to examine the political problems of the lyric as well as its possibilities.

If all of this sounds like an overstatement of lyric poetry's involvement in politics, that is largely because of poetry's reputation as a rarefied form that no one reads. But, as this book stresses, the idea of "lyric poetry" should not be conflated with the specific poems people write, read, or don't read. The *idea* of the lyric, whether people stop to read poems in public spaces, infuses daily life. As Mike Chasar observes, we are confronted with poetry all the time: its compression gives it an "easy mobility" relative to other types of literature, explaining its prominence in public spaces (from buses to museums) and across a range of media, including greeting cards, promotional materials, television, and Twitter.[35] Lyric poetry's portability is not just a matter of space but also of time. As Jahan Ramazani argues, poetry's temporality sets it apart from other genres; its "long-memoried" forms distinguish it from the swift, disposable time of the news—a contrast that explains much about its presence in the media coverage and ephemeral print culture of space exploration.[36] With its scalar compression in space and time, the lyric is an economical way to rhetorically immortalize and exalt fleeting national feats, from moon landings to shuttle launches.

The lyrical is the domain of special and weighty occasions, meant to bind its listeners together in mourning or in celebration. And indeed, many recent accounts of the lyric stress its communal and collective, rather than merely individual, dimensions. Bonnie Costello's *The Plural of Us* investigates the predominance of plural pronouns in contemporary poetry, especially in the shadow of W. H. Auden's often species-spanning "we."[37] Walt Hunter's *Forms of a World* investigates relationships between poetic form, human collectives, and global culture to argue for lyric poetry as a global, plural form.[38]

---

[35] Mike Chasar, *Poetry Unbound: Poems and New Media from the Magic Lantern to Instagram* (New York: Columbia University Press, 2020), 4.

[36] Jahan Ramazani, *Poetry and Its Others: News, Prayer, Song, and the Dialogue of Genres* (Chicago, IL: University of Chicago Press, 2014), 103, 67.

[37] Bonnie Costello, *The Plural of Us: Poetry and Community in Auden and Others* (Princeton, NJ: Princeton University Press, 2017).

[38] Walt Hunter, *Forms of a World: Contemporary Poetry and the Making of Globalization* (New York: Fordham University Press, 2019).

Moreover, critics including Culler and Barbara Johnson have noted how the lyric's staple rhetorical device of apostrophe is a fundamentally relational figure, positing and rearranging connections between subjects and objects.[39] Recently, Margaret Ronda has extended lyric apostrophe into theorizing relations not only between humans but also between humans and nonhuman others in the species-spanning Anthropocene.[40] In these critical accounts of the lyric "we," lyric collectivity is almost always desirable. The plural lyric would seem to free itself from what Gillian White describes as "lyric shame," or "shame experienced in identifications with modes of reading and writing understood to be lyric" tied to aesthetic conservatism and privileged apoliticism.[41] My contention is a bit different. Indeed, lyric poetry *does* involve the political and collective—but this plurality carries its own burdens. In the context of the postwar U.S., for instance, lyric rhetoric has galvanized collectivist expansionist fantasies. Sometimes, a lyric "we" is a single voice (usually white and male) masquerading as a human collective. On the flip side, if a lyric "we" writes or is read as the whole of a marginalized identity group, the lyric collective can threaten their claims to individuality and subjecthood, as I take up in ensuing chapters. One contribution of this book is to complicate the idea that the lyric's collective and communal dimensions are necessarily more desirable than its personal and individual ones, even in a plural age of globalization and planetary crisis.

I.2 The Lyric and the Cosmic Scale

The lyric might be thought of not so much as the realm of the individual nor the realm of the collective but, rather, as the realm of relational scales, all with their own ethical conundrums, from the tininess of an "I" to the largeness of national, global, planetary, and even cosmic spheres. The lyric and astronomy share a preoccupation with scale and the relative size of the human in the cosmos, making astronomical metaphors especially alluring for poets concerned with scale. At the most basic level, interpreting the night sky and interpreting a poem rely on scale. Scale is also a musical term, and one associated with the pitch intervals Pythagoras discovered that ancient Greek cosmology

---

[39] Culler, *Theory of the Lyric*, especially 242; Barbara Johnson brings apostrophe into the realm of political debate in "Apostrophe, Animation, Abortion," *Diacritics* 16:1 (Spring 1986): 28–47.

[40] Margaret Ronda, *Remainders: American Poetry at Nature's End* (Stanford, CA: Stanford University Press, 2018), 113–28.

[41] Gillian White, *Lyric Shame: The "Lyric" Subject of Contemporary American Poetry* (Cambridge, MA: Harvard University Press, 2014), 2.

mapped onto the celestial spheres. For lyric poetry, scale is both a matter of reading practices and of formal properties. We might think of the language of "close reading" (perhaps most intensively exercised on lyric poems) as well as Franco Moretti's corrective "distant reading."[42] Close reading and the neighboring practice of lyricization are processes of shrinking down, even as doing so opens up vast interpretive possibilities. Lyric reading aims, as Donne puts it in "The Good Morrow," to "make one little room an everywhere" ("room," the translation of the Italian word "stanza," puns on poetry's paradoxical compression and expansiveness).[43] The question of scale in lyric poetry is not just one of reading practices; scale also dictates many formal elements of the lyric. Emily Leithauser argues that lyric poetry "has always been about measurement and scale." The scalar orientation of poetry is most obvious in prosody, from meter to the relative scales of syllable, foot, line, stanza, and poem. This principle is less obviously but just as importantly true of figures; "Crudely speaking," writes Leithauser, "metaphor might be described as x = y," and its cousins metonymy and synecdoche rely on ideas of proximity and ratio.[44] Lyric address, too, hinges on and investigates scale, the relative distance, size, and weight of an "I" to a "thou." Lyric often amplifies the human subject by addressing a universe that outscales it, as when Tracy K. Smith measures her father against the cosmos, "the white cloud of his hair / In the distance like an eternity."[45] In other instances, the "I" addresses a "you" to modulate intimacy and distance. W. H. Auden, for instance, frequently uses lyric compression to invoke astronomical expansiveness through the relationship between an "I" and a "thou." In "The More Loving One," he describes the relative affection between himself and his lover through the metaphor of human relationships with the stars:

> How should we like it were stars to burn
> With a passion for us we could not return?
> If equal affection cannot be,
> Let the more loving one be me.[46]

---

[42] Franco Moretti, *Distant Reading* (New York: Verso, 2013). Moretti discusses the limitations of close reading, especially its perpetuation of a small canon, on pp. 48–9. He takes up the problem of lyric poetry's apparent requirement of close reading approaches on pp. 109–11.
[43] Donne, "The Good Morrow," 60.
[44] Emily Leithauser, "Remembering Poetry: Figures of Scale in the Postwar Anglophone Lyric," dissertation, Emory University, May 2016, 3.
[45] Tracy K. Smith, *Life on Mars* (Minneapolis: Graywolf Press, 2011), 34.
[46] W. H. Auden, "The More Loving One," in *Collected Poems*, ed. Edward Mendelson (New York: Random House, 2007), 582.

The poem investigates scale, from terms of measurement like "equal" and "more," to the physical distance of stars from the speaker, to the hyperbole of the metaphor itself, the ludicrousness of comparing disproportional love in a relationship to the nuclear fusion of stars that do not care about humans. The work of Smith, Auden, and others taken up in this book exemplifies how the postwar lyric draws heavily from astronomy and cosmology to investigate the scale-inflected ethics of belonging and relationality.

Lyric poetry, scale, and astronomy also intertwine in conceptions of the sublime. While Burke and Kant are the best-known visionaries of the sublime, a first-century CE text attributed to Longinus first articulates the concept at the meeting point of lyric poetry, scale, and *cosmos*.[47] Like the Greek gods tearing apart the harmonious universe in their rage, Longinus writes, lyric poetry captures tumultuous and conflicting emotions that tear apart the self. The sublime in epic concerns the polis and the external world; the sublime in lyric tears apart the internal realm. By way of example, he turns to Sappho's Fragment 31, in which the distressed speaker performs a wedding poem for a man and a woman whom she desires. In the fragment, the speaker enviously describes how the man manages, in the powerful ways of the gods, to remain unaffected by the beauty of the poem's beloved addressee, while she finds herself in "a cold sweat," tremulous, and feeling "nearly / to have died." Longinus calls this passage "sublime" in its presentation of a fragmented and contradictory self in turmoil; the speaker experiences her emotions as alien things that overwhelm her sense of bounded subjecthood as "she is assailed…by a tumult of different emotions" that undo her.[48] The first critical articulation of the sublime, then, locates it at the contact point of the lyric and the cosmos.

Burke's and Kant's formulations of the sublime, drivers of the Romantic and post-Romantic lyric, extend Longinus's emphasis on astronomy and subjectivity. Burke and Kant discussed the sublime just as developments in astronomy (some of which were theorized by Kant himself) increasingly revealed an incomprehensible cosmos. In his *Natural History and Theory of the Heavens* (1755), published two years prior to Burke's *A Philosophical Enquiry into the Origin of Our Ideas of the Sublime and Beautiful* (1757), Kant proposed the "nebular hypothesis," his more or less correct idea that the solar system formed from nebulous gas and dust. Using contemporary observations of other distant nebulae, Kant concluded that there must be many solar

---

[47] Longinus, *On the Sublime*, trans. H. L. Havell (London: Macmillan, 1890), X, 3.
[48] Longinus, *On the Sublime*, X, 23.

systems and galaxies, making ours just one of innumerable "island universes." While Kant got a lot of the technical details wrong, his work registered the expanding scale and intensifying unfathomability of a dynamic cosmos.[49] Burke and Kant draw frequently on cosmic wonder to describe the sublime. For Burke, extreme scales beyond human apprehension—both "greatness of dimension" and "the last degree of littleness"—produce the "astonishment" associated with the sublime.[50] If a situation is terrifying because it directly threatens to cause us harm, we experience only horror. But if the situation overwhelms or awes us without endangering us, we experience the sublime. If we're confronted with objects that far exceed human dimensions, such as a towering mountain or "the starry heaven," the experience is so incomprehensible that it shatters our rational faculties and our coherent sense of self breaks down.[51] Kant takes Burke's account of the subject undone by the sublime and turns it on its head: for Kant, contemplation of extreme scales extends subjectivity across the universe in an encounter with the infinite potential of one's own rational mind. Kant finds the sublime in a kind of cognitive hiccup, our very failure to comprehend the boundless universe. Scaling out from the example of a tree to a mountain to the Earth's diameter to the Milky Way, Kant proposes, "the immense multitude of such Milky Way systems, called nebulous stars, which presumably form another such system among themselves, do not lead us to expect any boundaries here," which "presents to us...our imagination, in all its boundlessness."[52] While Kant argues that no work of art can in itself produce sublime effects, he ranks poetry as the closest because, as when we try to comprehend the size of the Milky Way, poetry "expands the mind...it sets the imagination free" in its "unlimited variety of possible forms."[53]

Anna Barbauld's proto-Romantic 1773 poem "A Summer Evening's Meditation" richly integrates Kant's and Burke's cosmic sublime in the context of eighteenth-century discussions of nation building and critiques of Britain's expanding empire that would soon give way to the American one. The speaker wonders what drives her imagination "To the dread confines of eternal night, /

---

[49] John North offers an excellent overview of the developments in astronomy in this period, particularly of William Herschel's discoveries and catalogues of multiple nebulae that paved the way for nineteenth-century astronomy and Kant's imaginative misunderstanding of astronomer Thomas Wright's model of the Milky Way (*Norton History of Astronomy*, 398–409).
[50] Edmund Burke, *A Philosophical Enquiry into the Origin of Our Ideas of the Sublime and Beautiful*, ed. Adam Phillips (Oxford: Oxford University Press, 1998), 58.
[51] Burke, *A Philosophical Enquiry*, 66.
[52] Immanuel Kant, *Critique of Judgment*, trans. Werner Pluhar (Indianapolis, IN: Hackett, 1987), 113–14.
[53] Kant, *Critique of Judgment*, 196.

To solitudes of vast unpeopled space" of "embryo systems and unkindled suns," a nod to Kant's nebular hypothesis. She concludes that the enticing terror of the unfathomable draws her mind to cosmic contemplation. This activity will wreck her sense of bounded subjectivity even as her rational mind expands in its confrontation with unfathomable mysteries: she predicts a time when cosmic "splendours" will overwhelm her senses and "Unlock the glories of the world unknown."[54] Her later poem "Eighteen Hundred and Eleven" moves from cosmic unfathomability to a critique of a devolving England, "Where Power is seated, and where Science reigns." The poem condemns Britain's participation in the Napoleonic Wars, using scale-rich geographical and temporal metaphors to suggest that Britain has exceeded the reach of its power. Its attempts at global mastery, she predicts, will soon bring on another "Night, Gothic night," as Britain's power flags and the United States' influence grows.[55] Indeed, the U.S. would soon take up the mantle from Britain to articulate its own power in a cosmic-sublime register.

I.3 The American Lyric in Space

"The American bards...shall be *Kosmos*," Walt Whitman declared in the preface to *Leaves of Grass*, the urtext of American poetry.[56] As Ezra Pound put it, not only is Whitman "America's poet.... He *is* America."[57] In the story of Whitman, the poet himself, the U.S., and the cosmic all become lyrically enmeshed; Whitman wrote that "the United States themselves are essentially the greatest poem," their distinct parts merging into a harmonious whole (*CP*, 741). Whitman, *the* quintessential American poet, ostensibly healed the U.S. after the Civil War through his appeals to a repaired cosmic whole in a powerful lyric voice that was at once individual and collective. In "Song of Myself," Whitman imports epic expansiveness into the lyric mode, rerouting Virgil's opening of the *Aeneid* ("Of arms and the man I sing") into the self: "I celebrate myself, and sing myself, / And what I assume you shall assume, / For every atom belonging to me as good belongs to you" (*CP*, 63). These lines invert the scale of epic and lyric. Typically, the epic combines many distinct

---

[54] Anna Laetitia Barbauld, "A Summer Evening's Meditation," in *The Poems of Anna Letitia Barbauld* (Athens: University of Georgia Press, 1994), 83–4.
[55] Barbauld, "Eighteen Hundred and Eleven," in *The Poems of Anna Letitia Barbauld*, 156.
[56] Walt Whitman, *The Complete Poems*, ed. Francis Murphy (New York: Penguin, 2004), 750. Further references are noted in the text as *CP*.
[57] Ezra Pound, "What I Feel About Walt Whitman," in *Whitman*, ed. Roy Harvey Pearce (Englewood Cliffs, NJ: Prentice Hall, 1962), 8.

voices into the story of a single heroic empire or place; here, Whitman expands the individual "I" to contain "multitudes," as the tininess of the atom is a building block of expansive cosmic connections that are mirrored in the signature lengthy Whitmanian line built on tiny blocks of sound. "Song of Myself" is the blueprint for the American lyric, wherein a speaker (implicitly white and male) extends American individualism to speak prophetically on behalf of mankind. Richard Gray traces this impulse "back to the structure of feeling that Tocqueville perceived, a structure that has as its keystone the idea of the individual, the simple, separate self."[58] Whitman takes the American fascination with the "I" and welds it to the lyric "I" in ways that still reverberate through debates about poetic form and politics beyond U.S. borders, as I take up through the work of Agha Shahid Ali and Seamus Heaney in Chapters 3 and 4. Whitman uses cosmic idioms to inflate the scale of his "I" and project it across, even *as*, the universe, in a move that continues to haunt American poetry along its transnational routes.

In his declaration that "The American bards...shall be *Kosmos*," Whitman echoes Alexander von Humboldt, the Prussian scientist and polymath who revived the ancient Greek concept of *kosmos* following his five-year journey in the U.S., where he shaped nineteenth-century American thought as profoundly as Charles Darwin. Excavating Humboldt's largely forgotten influence in and beyond nineteenth-century America, Laura Dassow Walls uncovers how Humboldt's time in the U.S. suggested to him "that the physical universe and the human mind were integrated halves of a single whole, and he named that whole Cosmos."[59] In his poem "Kosmos," Whitman translates Humboldt's theory into a vision for American unity captured at the level of poetic form:

> The theory of a city, a poem, and of the large politics of
>   these States;
> Who believes not only in our globe with its sun and moon,
>   but in other globes with their suns and moons...
> The past, the future, dwelling there, like space, inseparable
>   together.
>
>   (*CP*, 413)

---

[58] Richard Gray, *A History of American Poetry* (Malden, MA: Wiley Blackwell, 2015), 9. The avant-garde tradition in particular has scrutinized the way the lyric self represents the liberal democratic American self and sustains capitalist ideologies. See, for instance, Sacvan Bercovitch, *The American Jeremiad* (Madison: University of Wisconsin Press, 1978), 176–210. Yet the avant-garde tradition has been overwhelmingly white, and in fact Whitman's aesthetics paved the way for its developments.

[59] Laura Dassow Walls, *Passage to Cosmos: Alexander von Humboldt and the Shaping of America* (Chicago, IL: University of Chicago Press, 2009), ix.

Whitman combines short phrases into sprawling lines, invoking the one that makes up the many and indeed democracy through the cosmic harmony of the poem's mode of combining the granular into a whole. In its time, Humboldt's work was recognized as "poetic" for its investment in harmony and design. Astronomer Maria Mitchell remarked after meeting Humboldt that "science... is not all mathematics, nor all logic, but it is somewhat beauty and poetry."[6]

Whitman invented the American lyric in a sublime mode just as American intellectual identity was being articulated at the juncture of the lyrical, the national, and the cosmic. In the nineteenth-century U.S., astronomy was melded onto the sublime in its expansion of American exceptionalism; indeed, astronomy was known as the "sublime science."[60] Ralph Waldo Emerson's *The American Scholar* pairs poetry and cosmic sublimity to consecrate a new American tradition. Aligned with the American Revolution-era metaphor of the new "Constellation" of the Republic, Emerson declares, "Our day of dependence, our long apprenticeship to the learning of other lands, draws to a close.... Who can doubt that poetry will revive and lead in a new age, as the star in the constellation Harp, which now flames in our zenith, astronomers announce, shall one day be the pole-star for a thousand years?"[61] Drawing on the powerful symbolism of the cosmically ordained American constellation, Emerson immortalizes the "new" land of America and its "new" intellectual tradition through the gravitas and long temporality of poetry, drawing on the lyric's purchase on the eternal and the cosmic. As Barry Ahearn has explained, "during the late 1830s and early 1840s 'lighthouses of the sky' began to dot the landscape of the United States. It would then be possible for the public to take pride in their nation's explorations of the heavens, and to hear lectures from American astronomers."[62] In this period, Whitman attended the New York lectures of astronomer Ormsby MacKnight Mitchel, who would go on to be known in the Union Army as "Old Stars." His poems record how nineteenth-century astronomy opened up expansive possibilities for American identity.

---

[60] Brett Zimmerman, "Nineteenth-Century American Astronomy and the Sublime," in *Journal of the Royal Astronomical Society of Canada* 97:3 (2003): 120–3, here 120.

[61] Ralph Waldo Emerson, "The American Scholar: An Oration, Delivered Before the Phi Beta Kappa Society, at Cambridge, August 31, 1837," in *The Collected Works of Ralph Waldo Emerson, Volume I*, ed. Robert E. Spiller (Cambridge, MA: Harvard University Press, 1971), 52–70, here 52.

[62] Barry Ahearn, *Pound, Frost, Moore, and Poetic Precision: Science in Modernist American Poetry* (New York: Palgrave Macmillan, 2020), 26.

Whitman's "When I Heard the Learn'd Astronomer," collected in his Civil War volume *Drum-Taps* (1865), is his best-known and most misread astronomical poem. In it, the lyric speaker apparently wanders away from an astronomer's dull lecture to enjoy a mystical communion with the night sky through poetic transcendence:

> When I heard the learn'd astronomer,
> When the proofs, the figures, were ranged in columns before me,
> When I was shown the charts and diagrams, to add, divide,
>   and measure them,
> When I sitting heard the astronomer where he lectured with
>   much applause in the lecture-room,
> How soon unaccountable I became tired and sick,
> Till rising and gliding out I wander'd off by myself,
> In the mystical moist night-air, and from time to time,
> Look'd up in perfect silence at the stars.
>
> <div align="right">(<i>CP</i>, 298)</div>

The poem's situation seems straightforward: in a familiar Romantic move, a lyric "I" rejects science for lyrical emotion. But the poem does not actually endorse the division of science and poetry. We can sense this first on the level of form. The poem encodes preoccupation with scale and measurement in its lines, which become longer and more laborious, like the astronomers' proofs, as the lecture drags on. Whitman then makes a pun on metrical versus free verse with the phrase "How soon unaccountable" that begins an unwieldy free verse line. The poem grows more metrically regular from that moment, until the final line settles into iambic pentameter, recalling the cosmic harmony of Milton's "Their starry dance in numbers that compute." The line "Look'd up in perfect silence at the stars," rendered in perfect iambic pentameter, records the perfection of the poem bolstered by the perfection of the cosmos. Indeed, as Whitman writes in the preface to *Leaves of Grass*, "the anatomist, chemist, astronomer, geologist, phrenologist, spiritualist, mathematician, historian and lexicographer, are not poets, but they are the lawgivers of poets, and their construction underlies the structure of every perfect poem" (*CP*, 751). The very science Whitman seems to denounce in favor of personal contemplation of the night sky actually informs both the meter and the operative metaphor at the end of the poem. Folsom notes that "the phrase from time to time...signals one of the newly formulated concepts that the astronomer would have explained in his lecture: that when we look at the stars, we are not only looking across vast distances of space, but vast distances of time as well."[63] Whitman's poem seems to put poetry and astronomy in opposition to one another, only to suggest the need for their union to achieve a sense of political and aesthetic harmony.

The Civil War context comes through in the poem's attention to harmony. As Folsom writes, "While the poem's subject is obviously not the Civil War, the tenor of the war times is nonetheless reflected in the speaker's desire to escape a place of fragmentation (where the unified cosmos is broken down and divided into "columns") and to regain a sense of wholeness."[64] Humboldt's *Kosmos* reverberates through Whitman's response to the Civil War. The poem thematically divides science and poetry, like the Confederacy and the Union, only to argue for their synthesis. In unifying science and poetry in this way, Whitman evokes a unified nation and cosmos fused through a lyric "I" that swells itself to the size of the universe.

Yet another architect of American poetry and astronomy enthusiast, Robert Frost also draws on astronomy to measure human scales against cosmic ones. "On Looking Up By Chance at the Constellations" concludes that although the cosmos is dynamic, humans cannot meaningfully register its changes and might as well rest assured in its apparent immobility: "Still it wouldn't reward the watcher to stay awake / In hopes of seeing the calm of heaven break / On his particular time and personal sight."[65] The poem stays anchored in limited human perception of cosmic vastness. Other astronomical poems by Frost project the "I" over the cosmos in a Whitmanian mode. "The Gift Outright," which Frost recited at John F. Kennedy's 1961 inauguration in the nascent days of the space race, draws on the expansive rhetoric of manifest destiny through the lyric "we" of the nation: "The land was ours before we were the land's" (424–5). In his astronomical poem "Desert Spaces," Frost's "I" projects itself into the same desert landscapes as the national "we" of "The Gift Outright." He contrasts the terrifying void of "empty spaces / Between stars" with the greater terror of empty places within the self: "I have it in me so much nearer home / To scare myself with my own desert places" (296). Through extreme scalar shifts, the poem communicates the interstellar chill of a mind rendered as at once familiar and uninterpretable. "Desert Places" precedes by almost three decades Kennedy's movement of manifest destiny from the "empty spaces" of the American frontier

---

[63] Ed Folsom, "When I Heard the Learn'd Astronomer," in J. R. LeMaster and Donald D. Kummings, eds., *Walt Whitman: An Encyclopedia* (New York: Garland Publishing, 1998), 769.

[64] Folsom, "When I Heard the Learn'd Astronomer," 769.

[65] Robert Frost, *The Poetry of Robert Frost: The Collected Poems*, ed. Edward. Connery Lathem (New York: Henry Holt, 2002), 268. Further references are noted in the text.

to the final frontier of extraterrestrial space and the new desert world of the moon. Yet it demonstrates an American tendency to project the "I," whether outward or inward, into spaces imagined as empty.

Ezra Pound and T. S. Eliot took the universalist, expansive strain in American poetry and welded it to scientific discourse in ways that would be indelible. Pound specifically reached for a "scientific spirit" in his own writings.[66] The first line of *ABC of Reading* declares that his era is one of "science and abundance," a description that invokes vast spaces and material plenitude.[67] Lisa Steinman argues that scientific discourse for modernist poets was often rooted in national identity, as these poets worried about the U.S.'s marginalization of poetry and responded by suggesting poetry itself as an innovative American technology. Specifically, many modernist poets linked the poetic imagination to the visionary work of the New Physics associated with Einstein.[68] Both Pound and Eliot attempted to make a science out of lyric subjectivity. As Pound put it, "The arts, literature, poesy, are a science, just as chemistry is a science. Their subject is man, mankind and the individual."[69] Following Pound, Eliot influentially universalized a lyric "I" by imbuing it with scientific authority. In "Tradition and the Individual Talent," Eliot famously remarks, "It is in this depersonalization that art may be said to approach the condition of science."[70] Influenced by Pound and Eliot, the New Criticism modeled itself on scientific discourse and methods. The poets in this book share a cluster of interests around cosmology and poetic form shaped by the New Criticism, which amplified the idea of lyric universality through scientific rhetoric. In fact, as Peter Middleton demonstrates in *Physics Envy*, the New Criticism arose in part as a threatened reaction to the perceived "epistemic authority of the natural sciences" at mid-century, an era in which "to be an American was indeed to be scientific."[71] As Middleton has shown, this relationship drove much mid-century Cold War poetry as New Critical methodologies dominated in universities. The postwar era, then, saw science become not just a subject and method for poetry but also an ideological force for it to incorporate and resist.

---

[66] Ezra Pound, *Selected Prose, 1909–1965*, ed. William Cookson (London: Faber & Faber, 1973), 28.
[67] Ezra Pound, *ABC of Reading* (London: Faber & Faber, 1991), 17.
[68] Lisa Steinman, *Made in America: Science, Technology, and American Modernist Poets* (New Haven, CT: Yale University Press, 1987).
[69] Pound, *Selected Prose*, 28.
[70] T. S. Eliot, "Tradition and the Individual Talent," in *The Sacred Wood: Essays on Poetry and Criticism* (1920; repr., London: Faber & Faber, 1997), 47.
[71] Peter Middleton, *Physics Envy: American Poetry and Science in the Cold War and After* (Chicago, IL: University of Chicago Press, 2015), 49, 8.

I.4 The Lyric in Cold War Culture

While Middleton focuses on physics as the most rhetorically powerful of mid-century sciences, astronomy, due to its proximity to geography, better captures the interplay of epistemological and geographical mastery that drove the Cold War as the U.S. and the USSR battled for control of the globe. The space race came out of a postwar fusion of scientific and military power. As Sarah Daw has put it, "collective media endorsement of Cold War science was often couched in the language of conquest," especially conquest over the natural world.[72] The Soviet launch of Sputnik 1 in 1957 threatened Americans' scientific superiority and its attendant promise of planetary domination. Sputnik led to the founding of NASA in 1958 and the National Defense Education Act, which funneled over one billion dollars into science and engineering education the same year. The act revived astronomy education in the U.S. after a sixty-year hiatus, placing astronomy in a curriculum aimed at extending individualist values.[73] The Second World War had driven advances in astrophysics like rocketry, radar, and X-rays that revolutionized study of the universe and propelled the U.S.'s rise to global superpower, partly through former Nazi rocket scientists like Wernher von Braun brought to work on the space program through Operation Paperclip. With new visibility in schools, proliferating planetariums, and cosmic-themed merchandise, astronomy amplified U.S. power on the Cold War stage through both scientific advances and powerful symbolism. Early in the space race, American mastery of the universe was both a metaphor for mastery of Earth and a literal precondition for global domination. As Kennedy put it during his presidential campaign in 1960, "Control of space will be decided in the next decade. If the Soviets control space they can control earth, as in past centuries the nation that controlled the seas dominated the continents."[74] The next year, then Vice President Lyndon Johnson advised Kennedy, "…failure to master space means being second-best in the crucial arena of our Cold War world. In the eyes of the world, first in space means first, period; second in space is second in

---

[72] Sarah Daw, *Writing Nature in Cold War American Literature* (Edinburgh, UK: Edinburgh University Press, 2018), 3.
[73] For a history of astronomy education in the U.S., see D. B. Hoff, "History of the Teaching of Astronomy in American High Schools," in J. M. Pasachoff and J. R. Percy, eds., *Proceedings of IAU Colloq. 105, held in Williamstown, MA, 27–30 July 1988* (Cambridge: Cambridge University Press, 1990), 249–53.
[74] John F. Kennedy, "If the Soviets Control Space, They Can Control Earth," *Missiles and Rockets* 7:15 (10 Oct. 1960): 12–13, here 12.

everything."[75] Kennedy's "we choose to go to the moon" speech the following year blended the language of mastering space with manifest destiny, describing the moon as "the furthest outpost on the new frontier of science and space."[76] Space race rhetoric at its peak conflated epistemological and military mastery of the universe, as expansion in the name of science eclipsed the imperialist origins and uses of space science. The very phrase space *age* naturalizes space exploration as an evolutionary inevitability, even as the rapid technological innovations associated with it suggested an American capacity to accelerate evolution's long timescales.

Because the Soviets demonstrated technical superiority, beating the U.S. into space at every juncture until the lunar orbit of Apollo 8 and the moon landing itself, the U.S. was left with symbolic, romantic narratives about space exploration. At battle in space for control of the planet, therefore, the U.S. turned to lyric, even sending Thomas Bergin's 1961 poem "For a Space Explorer" up on a Navy satellite.[77] Apollo-era institutions drew on the ceremonial, temporal, and quasi-religious dimensions of lyric to present the nationalism that drove space exploration as a universal, inevitable, even divinely backed, event. Even the name of the program, "Project Apollo," played into this strategy. The moniker, announced publicly in July 1960, was chosen by Abe Silverstein, NASA's Director of Space Flight and Development. The explanation given in the 1976 official publication *Origins of NASA Names* focuses on the "attractive connotations" of the Greek god: "Apollo was god of archery, prophecy, poetry, and music, and most significantly he was god of the Sun."[78] While the account given by the Glenn Research Center bio on Silverstein makes the choice seem casual—"Silverstein chose the name 'Apollo' after perusing a book of mythology at home one evening in 1960" because it seemed "appropriate to the grand scale of the proposed program'"—the poetry-heavy "What's In A Name?" highlights the deliberate lyrical

---

[75] L. B. Johnson, "Evaluation of Space Program," in John M. Logsdon., ed., *Exploring the Unknown: Selected Documents in the History of the U.S. Civil Space Program* (Washington, D.C., 2004).

[76] John F. Kennedy, Moon Speech, Rice Stadium, Rice University, 12 Sept. 1962, https://er.jsc.nasa.gov/seh/ricetalk.htm.

[77] Philip Leonard, *Orbital Poetics: Literature, Theory, World* (London: Bloomsbury Academic, 2019), 50.

[78] Helen T. Wells, Susan H. Whiteley, and Carrie E. Karegeannes, *Origins of NASA Names*, The NASA History Series (Washington, D.C.: National Aeronautics and Space Administration, 1976), 99, https://history.nasa.gov/SP-4402.pdf. Apollo was not the only poetic name NASA chose for its projects. Silverstein had previously named the Project Mercury, NASA's first crewed program, after the Roman messenger of the gods, and the name Orpheus was considered for what ultimately became Project Gemini (106, 104). Even the term *astronaut* "followed the semantic tradition begun with 'Argonauts,' the legendary Greeks traveled far and wide in search of the Golden Fleece," and, as the authors note, was first "found in the writings of French poet Cyrano de Bergerac" (107, 200).

impulse behind the choice: "A rose by any other name may smell as sweet. But would a mission to the moon named 'Pegasus' have the same appeal as Apollo? Thanks to Dr. Abe Silverstein... we'll never know."[79]

Helene Knox discusses the government's yearning for space poetry in a piece commemorating twenty-five years of NASA space exploration: "From the beginning of the program, many people longed for an eloquent communication from spacefarers, something more than 'Everything is A-OK—the view is really great up here!' Among the grumblings, the idea surfaced early that NASA ought to send a poet into space."[80] The desire to send up a poet registered the worry that the technological and bureaucratic realities of space exploration had killed the "poetry" of space. As a result, many texts on poetry and the space age were published in the 1960s to bring back the romance of space and, as Laurence Goldstein has put it, "smooth the way for NASA."[81] For instance, Arthur C. Clarke's essay "Astronautics and Poetry" was reprinted in *The Coming of the Space Age* (1967), a collection of essays meant to lend intellectual and poetic gravitas to NASA's exploits. First published by the British Interplanetary Society in 1947, the essay positions the space age within a poetic-astronomical tradition stretching from Lucretius to Shelley, Tennyson, and Whitman. Clarke ends with a rebuttal of the Roman astrologer-poet Manilius, who, in A. E. Housman's translation, declared, "Content you with the mimic heaven, / And on the earth remain." "Excellent advice, perhaps," Clarke writes, "but we shall not take it."[82] The essay's republication in a collection meant to lend credence to NASA's adventures imbues Anglo-American space exploration with the gravitas of a long poetic tradition.

NASA promotional materials frequently recycle lines from Alfred, Lord Tennyson (an amateur astronomer during the revolution in modern astronomy in Victorian England) and T. S. Eliot. Despite these poets' scientific interests, promotional materials mine their work for expansionist language. Rather

---

[79] Emily Kennard, "What's In A Name?", Glenn Research Center, 15 May 2009, https://www.nasa.gov/centers/glenn/about/history/silverstein_feature.html. Of course, Pegasus would have fit NASA's criteria just as well. Depending on what one reads, Pegasus was raised by the Muses, created the springs from which the Muses sprung, and/or created one particular spring (the Hippocrene) sacred to the muses. Pegasus is especially associated with poets and currently adorns the logo for the Poetry Foundation.

[80] Helene Knox, "Space Poems: Close Encounters Between the Lyric Imagination and 25 Years of NASA Space Exploration," in *Lunar Bases and Space Activities of the 21st Century*, ed. W. W. Mendell (Houston, TX: Lunar and Planetary Institute, 1985), 771. Knox's piece was written for the Lunar and Planetary Institute, which supports NASA and planetary science.

[81] Laurence Goldstein, "'The End of All Our Exploring': The Moon Landing and Modern Poetry," *Michigan Quarterly Review* 18:2 (1979): 192–216, here 193.

[82] Arthur C. Clarke, "Astronautics and Poetry," in *The Coming of the Space Age: Famous Accounts of Man's Probing of the Universe* (New York: Meredith Press, 1967), 293–7, here 297.

than excerpting Tennyson's many poems about cosmic entropy, NASA usually selects lines from his poem of exploration and conquest "Ulysses": "To sail beyond the sunset, and the baths / Of all the western stars."[83] Eliot's "Little Gidding," which is not about astronomy at all, is another NASA favorite:

> We shall not cease from exploration
> And the end of all our exploring
> Will be to arrive where we started
> And know the place for the first time.[84]

Steven J. Dick, NASA's former Chief Historian (2003–9), argues that Eliot's "sentiment poetically captures one of the most important reasons we explore space. NASA's exploration of the universe, and that of other nations, reveals humanity's place in nature in the broadest possible sense."[85] For "humanity," however, read "England," the historical and aesthetic preoccupation of *Four Quartets*. By taking two poets who often universalize British and European culture through the language of human exploration, the U.S. project of space exploration reveals the imperialist principles underlying it through the very art meant to pacify and exalt its ventures.

The lyric has bolstered an American nationalist project to present space exploration as a natural outgrowth of the human condition. NASA's interest in poetry has ties to a public relations strategy of romanticizing space exploration. Since its founding in 1958, NASA, more than other government agencies, has depended on public support for its funding. Mark E. Byrnes observes that NASA has galvanized public support primarily through appeals to "nationalism, romanticism, and pragmatism." Byrnes notes that romantic rhetoric drove the age of Apollo by arguing that space exploration "allows humans to continue to explore, requires the efforts of heroic people, provides a variety of emotional rewards, and helps satisfy human curiosity." As Byrnes puts it, an important strain of romanticism in NASA image-making has been

---

[83] Examples of the uses of these lines are too numerous to count. As Goldstein puts it, "Hardly a book exists about space travel that does not quote Tennyson's 'Ulysses' on the need to seek a newer world" (194). Following the *Challenger* disaster in 1986 that effectively ended the first space age, the Challenger Campaign commissioned Robert A. Heinlein to write "A Letter to the American People." William H. Patterson suggests that this letter influenced Heinlein's book *To Sail Beyond the Sunset*, which takes its title from Tennyson's "Ulysses," which is "about picking up and moving on, as humankind has always rolled on." William H. Patterson, *Robert A. Heinlein: In Dialogue with His Century, Volume 2: The Man Who Learned Better, 1948–1988* (New York: Tor, 2016), 451.

[84] T. S. Eliot, "Little Gidding," from *Four Quartets*, in *T. S. Eliot: The Complete Poems and Plays 1909–1950* (Boston: Houghton Mifflin, 2014), 145.

[85] Steven J. Dick, "Why We Explore," 21 July 2005, https://www.nasa.gov/exploration/whyweexplore/Why_We_13.html.

"the idea that NASA helps fulfill the ingrained human urge, particularly strong in Americans, to explore the unknown and expand the frontier."[86] Lyric poetry and rhetoric, as part of this marketing strategy, have at times translated militarism into the soaring of the human spirit. Philip Leonard argues that, "As part of its efforts to disavow the idea that orbiting instruments and space exploration express a will to dominate from the heavens, NASA has continued to send poetry into space."[87] I investigate the bizarre practice of sending poems into space in the Coda.

Poets, for the most part, were not inspired by NASA's lyricism. The Aerospace Industries Association conducted a review of how space exploration had impacted cultural production and were especially disappointed to find "years of silence" from poets, leading them to bemoan how space exploration had not inspired more poetic creations.[88] "Mneh!" was Auden's uninspired reaction in his anti-NASA poem "Moon Landing."[89] The *New York Times* commissioned responses by poets in response to the Apollo 11 moon landing, which mostly registered displeasure. Babette Deutsch, whose poetry grows increasingly disenchanted with space exploration, asks in "To the Moon, 1969," "Are you a monster? / A noble being? Or simply a planet that men have, / almost casually, cheapened?"[90] June Jordan offered an even more searing critique in the *Times* spread, echoing many Black artists' and intellectuals' protest of the moon landing during the Civil Rights era: "I mean, brothers and sisters, have you ever heard of children—bankrupt, screaming—on the moon?"[91] Many other texts meant to popularize space exploration through poetry also registered poets' discontents. Robert Vas Dias's anthology *Inside Outer Space*, for instance, is an assortment of underwhelmed, dystopian poems mostly by white male leftish intellectuals from the Northeast who feared that technological and bureaucratic takeover would undermine their craft.[92] This underwhelmed response to space exploration caught space program popularizers off guard. As Laurence Goldstein writes in his history of moon landing poetry, "From all accounts the negative response of poets...took supporters of the space program by

---

[86] Mark E. Byrnes, *Politics and Space: Image Making by NASA* (Westport, CT: Praeger, 1994), 3, 48.
[87] Leonard, *Orbital Poetics*, 49–50.
[88] Quoted in Robert Poole, *Earthrise: How Man First Saw Earth* (New Haven: Yale University Press, 2009), 192.
[89] W. H. Auden, "Moon Landing," in *Collected Poems*, ed. Edward Mendelson (New York: Random House, 2007), 845.
[90] Babette Deutsch, "To the Moon, 1969," *New York Times*, 21 July 1969, 17.
[91] June Jordan, quoted in the *New York Times*, 21 July 1969, 17.
[92] Robert Vas Dias, *Inside Outer Space: New Poems of the Space Age* (New York: Anchor Books, 1970).

complete surprise. If there was one ally in the public they had counted on it was the visionary poets."[93] NASA commissioned two poets to be visionaries, Archibald MacLeish and James Dickey, whom I discuss in Chapter 1.

In addition to commissioning poets, NASA tried to enlist astronauts, urging them to be more lyrical about their experiences in space. Astronauts were frequently mocked in the press for delivering dull reports about their adventures in the wonders of space. Dismayed by astronauts' ineloquent descriptions, Kurt Vonnegut, Jr., created a 1972 PBS special, *Between Time and Timbuktu*, that followed the adventures of Stony Stevenson, the first poet in space.[94] Astronauts, for their parts, resisted the expectation to be poetic. Michael Collins of Apollo 11 complained, "We weren't trained to emote, we were trained to repress emotions. If they wanted an emotional press conference, for Christ's sake, they should have put together an Apollo crew of a philosopher, a priest, and a poet—not three test pilots."[95] (Collins added that this expressive crew would "kill themselves trying to fly the spacecraft.")[96] Frank Borman, crewman on the Apollo 8, expressed similar opinions about space and poetry. His mission resulted in the first widely distributed photograph of planet Earth taken from space, *Earthrise*. In a *This American Life* segment on the fiftieth anniversary of the *Earthrise* photograph, science reporter David Kestenbaum asked Borman, "Did you say, at some point, they should have sent a poet?" Borman replied, "No, I didn't—if I did, I didn't [mean it]—the last thing I would have wanted on our crew was a *poet*."[97] The device of the segment is that Borman, "the least complicated man" on the planet, is the furthest thing from an aesthete and is nonetheless moved, for perhaps the only time in his life, by seeing Earth from space. When Borman starts talking about the Earth, the podcast music becomes slow and sentimental, and Borman himself is moved: "The thing...that I [will] recall till the day I die, was the Earth, looking back at the Earth." As Chapter 1 will explore, the most "lyrical" cultural and aesthetic dimensions of space exploration have emerged around planet Earth, which becomes, as I will argue, a lyrical object in the late 1960s.

---

[93] Goldstein, "'The End of All Our Exploring'," 193.
[94] Matthew D. Tribbe discusses *Between Time and Timbuktu* in the context of NASA's Apollo program in *No Requiem for the Space Age: The Apollo Moon Landings and American Culture* (Oxford: Oxford University Press, 2014), 42.
[95] Quoted in Tribbe, *No Requiem for the Space Age*, 42.
[96] Frank White, *The Overview Effect: Space Exploration and Human Evolution* (Reston, VA: American Institute of Aeronautics and Astronautics, 2014), 37.
[97] David Kestenbaum, "So Over the Moon," in "655: The Not-So-Great Unknown," host Ira Glass, produced by WBEZ. *This American Life*, 24 Aug. 2018. Podcast, MP3 audio, https://www.thisamericanlife.org/655/the-not-so-great-unknown.

Gene Roddenberry's *Star Trek: The Original Series* (1966–9), which originally aired during the height of the space race, also uses lyric poetry to romanticize space travel and to make the show's adventures more highbrow. *Star Trek* often incorporates the lyric mode—especially through Romantic poetry and Shakespeare, who is typically quoted in snippets and presented in a Romantic vein—to signal its seriousness.[98] *Star Trek* presented itself as thoughtful political science fiction that challenged Cold War ideology by backing civil rights and criticizing the Vietnam War. But the starship *Enterprise*—an on-the-nose capitalist name that NASA used for a test craft in homage to the show—is also a vessel of Cold War values as it sails through "space: the final frontier." The ship's mission, described in its opening monologue, is "to boldly go where no man has gone before," a line that recycles language from both Captain Cook and a 1958 White House document produced to quell Sputnik panic called "Introduction to Outer Space."[99] Lyric in *Star Trek*, as for NASA, is meant to counter the technological and militaristic dimensions of spacefaring. Multiple episode titles draw from Romantic poetry, most obviously "Who Mourns for Adonais?" (1967) from Shelley (with an echo of Milton's "Lycidas"), and "Is There in Truth No Beauty?" (1968) from Keats. Other *Star Trek* poetry references frequently combine Greek aesthetics, sexual desire, and masculine conquest in a Romantic lyric vein. In "Whom Gods Destroy," the green-skinned inmate Marta claims that she wrote both Shakespeare's Sonnet 18 and a pastoral ballad by A. E. Housman before attempting to seduce Captain Kirk, who is described by another lover, in the Shakespearean episode "Wink of an Eye" (a phrase from *The Winter's Tale*) as "Caesar of the stars."[100] The example of plagiarized poetry in "Whom Gods Destroy" also recycles a frequent trope of mid-century television. As Chasar writes, "Forced to meet pressing and seemingly endless content demands, television writers faced the prospect that unless they relied on formulas or clichés or stole from other shows, they might run out of material. In one respect, the plagiarism plot motif provided those writers with an

---

[98] Most work on poetry in *Star Trek* focuses on Shakespeare. See, for instance, Mary Buhl Dutta, "'Very bad poetry, Captain': Shakespeare in *Star Trek*," *Extrapolation* 36:1 (1995): 38–45; and Paul A. Cantor, "Shakespeare in the Original Klingon: *Star Trek* and the End of History," *Perspectives on Political Science* 29:3 (2000): 158–66.
[99] The White House document describes "...the compelling urge of man to explore and to discover, the thrust of curiosity that leads men to try to go where no one has gone before." See President's Science Advisory Committee, "Introduction to Outer Space," 26 Mar. 1958, 2.
[100] *Star Trek: The Original Series*, season 3, episode 14, "Whom Gods Destroy," directed by Herb Wallerstein, written by Lee Erwin, featuring William Shatner, Leonard Nimoy, and DeForest Kelley, aired 3 Jan. 1969; and season 3, episode 11, "Wink of an Eye," directed by Jud Taylor, written by Arthur Heinemann, featuring William Shatner, Leonard Nimoy, and DeForest Kelley, aired 29 Nov. 1968.

opportunity to process this anxiety."[101] A *Macbeth*-themed Halloween episode does exactly this when the trochaic lines of three witch figures on a foggy planet lead Spock to declare, "Very bad poetry, Captain"—a tongue-in-cheek nod to the very bad episode.

The ostensibly emotionless scientist Spock, in fact, is the character most associated with lyric poetry. In "Charlie X," the episode's titular antagonist forces Spock to recite the opening lines of William Blake's "The Tyger," a poem that is appropriately demonic for the pointy-eared character. Spock is an Orphic figure, an alien associated with the underworld who even plays the Vulcan lyre. The scientist with a flair for poetry—"Poetry, Captain. Nonregulation," he compliments Kirk's heartfelt speech at the end of "This Side of Paradise"—brings together Vulcan and human halves, healing rifts between peoples, species, and planets by humanizing a literally non-human scientist through the lyrical. This mending of a painful rift between the sciences and the humanities is a staple of postwar American culture.

Whitman, NASA, and *Star Trek* all suggest a conflict between "astronomy" and "poetry" that must be healed for the good of American democracy. In 1959, C. P. Snow lamented this apparent rift between the humanities and the sciences in his lecture on the "two cultures" that could no longer communicate across a vast gulf.[102] Scott McLemee notes "the Cold War subtext (two cultures glowering across an abyss)" of Snow in his review of Tracy Daugherty's *Dante and the Early Astronomer*.[103] Repairing their rift, suggests 1960s space culture, would save the world. Harmonizing poetry and science became the cultural mission of astronomer Carl Sagan during the atomic age, as he repeatedly lyricized astronomy to American audiences in an attempt to avert nuclear disaster. Sagan's *Cosmos* teems with Cold War nuclear anxiety and lyrical prose. Michael Page specifically likens Sagan's style to Romantic poetry, writing that it "is fueled by the same visionary temperament that invests [Erasmus] Darwin's poetry."[104] Sagan loved poetry. He served on the English Ph.D. committee of poet Diane Ackerman, and she dedicates the

---

[101] Chasar, *Poetry Unbound*, 20. He adds, "the association of poetry and plagiarism via storylines in which characters are caught plagiarizing poems or fake being poets" is so engrained that "in the world of television, poetry now frequently functions as a type of clue, tell, sign, signal, or shorthand that, somewhere close by, some sort of plagiarism or equivalent act of fraudulence, fakery, or imposture is going on or just about to happen."

[102] C. P. Snow, *The Two Cultures and the Scientific Revolution* (Cambridge: Cambridge University Press, 1959).

[103] Scott McLemee, "Sunspots and Poetry," *Inside Higher Ed*, 7 June 2019, https://www.insidehighered.com/views/2019/06/07/review-tracy-daugherty-dante-and-early-astronomer-science-adventure-and-victorian.

[104] Michael Page, "The Darwin Before Darwin: Erasmus Darwin, Visionary Science, and Romantic Poetry," *Papers on Language and Literature* 41:2 (2005): 146–69, here 146.

poem "Venus" in *The Planets: Cosmic Pastoral* to Sagan.[105] Sagan's essay "Space, Time and the Poet," which he wrote for his high school newspaper, comments on astronomical poems by Tennyson, Poe, Milton, Helen Hunt Jackson, Frost, Eliot, and others. It ends, "After journeying through space over the galactic hub and through time to the terminus of our puny planet, we must be impressed with a feeling of Man's utter insignificance before the universe"—followed by suggestions of further space-themed poetry to read.[106] Sagan's attempts to harmonize science, poetry, and the Cold War globe bring to mind Whitman's astronomical poetry of the Civil War. Indeed, Walls traces the title and concept of Sagan's hit show *Cosmos* to Humboldt's *Kosmos*, pointing out that "His television series *Cosmos*—in which Sagan appeared to travel by spaceship to the farthest reaches of outer space and then return slowly to Earth—bore striking parallels to Humboldt's masterwork."[107] Sagan imagines reunifying a world split into realms of scientific and emotional responses, in order to repair not the North–South divide of the U.S. Civil War that preoccupied Whitman but instead the East–West bifurcation of the Cold War globe. When astrophysicist Neil deGrasse Tyson took over *Cosmos* and the role of the public astronomer, Whitman's "When I Heard the Learn'd Astronomer" became his go-to piece for public appearances, from Hayden Planetarium's *Dark Universe* show to popular science interviews.[108] It is, in fact, a staple of contemporary American culture. It features in two episodes of *Breaking Bad*.[109] It also featured in a Whitman-themed pop-up event of Verse and Universe, an annual charitable event that brings together poets and

[105] Diane Ackerman, *The Planets: Cosmic Pastoral* (New York: William Morrow & Co., 1976). Sagan sent one of Ackerman's poems to Timothy Leary in prison, included with a letter discussing the possibility of extraterrestrial life. Leary and Sagan were both political activists who opposed Reagan's Cold War policies, including the "Star Wars" program. See Lisa Rein and Michael Horowitz, "Inner Space and Outer Space: Carl Sagan's Letters to Timothy Leary (1974)," in the *Timothy Leary Archives*, http://www.timothylearyarchives.org/carl-sagans-letters-to-timothy-leary-1974/.
[106] Rob Casper, "Space, Time, and the Poet Sagan," in *From the Catbird Seat: Poetry & Literature at the Library of Congress*, 30 Jan. 2014, https://blogs.loc.gov/catbird/2014/01/space-time-and-the-poet-sagan/.
[107] Laura Dassow Walls, "O Pioneer," *American Scientist* 104:2 (March–April 2016): 118, https://www.americanscientist.org/article/o-pioneer.
[108] He read it, for instance, to promote Hayden Planetarium's fifth show, *Dark Universe* (2013–20), where the poem expresses the awe and wonder of cosmic scale and the immeasurability of cosmic phenomena like dark matter. See "'When I Heard the Learn'd Astronomer' with Neil deGrasse Tyson," 14 Nov. 2013, video, 0:55, https://www.amnh.org/exhibitions/permanent/hayden-planetarium/dark-universe. DeGrasse Tyson has read this poem in other public pedagogical contexts, including for an Art of Charm video: "Neil deGrasse Tyson Reads When I Heard The Learn'd Astronomer," *YouTube*, 6 June 2017, 1:08, https://www.youtube.com/watch?v=vfl-ACKmpK8.
[109] John Shiban, writer, *Breaking Bad*, Season 3, episode 6, "Sunset," directed by John Shiban, featuring Bryan Cranston, Aaron Paul, and David Costabile, aired 25 Apr. 2010. The poem appears again in Moira Walley-Beckett, writer, *Breaking Bad*, Season 4, episode 4, "Bullet Points," directed by Colin Bucksey, featuring Bryan Cranston, Aaron Paul, and David Costabile, aired 7 Aug. 2011.

astrophysicists to promote science funding and environmental stewardship.[110] The union of poetic and astronomical discourse lives deep in the American cultural imaginary.

I.5 The Lyric in the New Space Age

In the new space age of billionaire-driven space exploration, poets continue to write out of and against U.S. nationalism. Fatimah Asghar's "Pluto Shits on the Universe" uses metaphors of cosmic harmony and nationalism to argue for a new conception of the cosmos that reflects her experiences as a queer Kashmiri American woman. The persona poem assumes the voice of Pluto, demoted from planetary status in 2006 for its chaotic and irregular orbit. Like Pluto, Asghar suggests, she can "chaos like a motherfucker," breaking Western aesthetic-political orderly systems that do not know how to classify her or, perhaps, her poetry:

> I realigned the cosmos.
> I chaosed all the hell you have yet to feel. Now all your kids
> in the classrooms, they confused. All their clocks:
> wrong. They don't even know what the fuck to do.
> They gotta memorize new songs and shit. And the other
> planets, I fucked their orbits. I shook the sky.[111]

The songs Asghar references to help American schoolchildren learn the order of the planets were popularized during the Cold War, in the 1958 Education Defense Act's attempts to make U.S. citizens "masters of the universe." Asghar's poem creates its own rhythms and songs, realigning interlinked ideas of the cosmos and the lyric poem.

The year after the Space Force went into effect, Sumita Chakraborty's cosmic book *Arrow* (2020) debuted. "Marigolds," its opening poem, begins with a description of confessional poet Robert Lowell, only to subvert the confessional mode through cosmic inscrutability. The lyric "I" is impossible to chart as it slips across geographies from Boston to the deep Pacific Ocean and

---

[110] For an archive of these events since 2017, visit https://www.brainpickings.org/the-universe-in-verse/.
[111] Fatimah Asghar, "Pluto Shits on the Universe," *Poetry Magazine* (April 2015), https://www.poetryfoundation.org/poetrymagazine/poems/58056/pluto-shits-on-the-universe.

Malaysia Airlines Flight 370 to a moon of Saturn. The "I" contemplates precise distances that ought to overwhelm it: "Between the wine-colored hull of Ocean Shield and Enceladus / lies eight times the distance between Earth and the sun."[112] In the concluding couplet, the "I" projects itself far into space not in an act of conquest but, rather, in an act of slipping through dominant epistemic structures: "Worlds such as this were not thought possible to exist. / My lord, I aim a mile beyond the honeyed moon." The penultimate line, which recurs throughout *Arrow*, is taken from a 2014 NASA news story about the discovery of an exoplanet whose physical properties "confounded" astronomers, a planet that somehow maintained a rocky composition despite "enormous gravitational force" that should have resulted in a ballooning gas giant.[113] Yet, *Arrow* suggests, such worlds should not come as a surprise. Throughout the volume, the lyric "I," governed by the gravity of horrific violence from the domestic to the planetary, finds the existence of a supermassive rocky planet to be no surprise at all; she has always inhabited the kind of world that shocked these astronomers. The impossible world suggests not only devastation but also the act of lyric creation. Throughout "Marigolds," the apostrophic "O" invokes the shape of worlds and invites meditation on the act of lyric as world creation. As the poet mourns her sister's death—a devastating result of domestic, systemic, and biological violence—lyric itself becomes an otherworldly landscape of both grieving and making. Indeed, "poetry," the act of bringing into existence what didn't exist before, is a mode of conjuring worlds that "were not thought possible to exist." Much as the exoplanet defies assumptions in planetary science in its arrangement of matter, the poet creates unexpected, even mystifying poems out of inherited lyric tools like apostrophe.

The following chapters begin with the nearest world, planet Earth, and scale outward.

Chapter 1 argues that planet Earth became a lyrical object in the late 1960s, a phenomenon that has shaped contemporary understandings of the planet and the direction of contemporary lyric. In 1968, NASA's iconic photograph *Earthrise*, showing a vibrant planet Earth over a desolate moon, hit the public with revelatory force. This photograph and others like it have variously served as icons of the Anthropocene, militant artifacts of U.S. imperialism, and

---

[112] Sumita Chakraborty, "Marigolds," in *Arrow* (Farmington, ME: Alice James Books, 2020), 5.
[113] "Astronomers Confounded by Massive Rocky World," 2 June 2014, *NASA*, https://www.nasa.gov/ames/kepler/astronomers-confounded-by-massive-rocky-world.

maudlin portraits of a united humanity. Lyric has been *the* mode for contending with these competing visions of the whole Earth. *National Geographic, Life*, the *Times*, and the *New York Times* all distributed the first photographs of Earth from space with poetry; hundreds of poets around the world have written about them; and space historians and astronauts wax poetic about the view. To understand the phenomenon of the lyrical Earth, I turn to lyric figures—particularly apostrophe, graphically rendered in the O of Earth (see Section 1.2). Understanding "planet Earth" as a lyrical construction reveals it to be at once an American Cold War artifact and a strange, unassimilable otherworld. Chapter 2, "'Galaxies of Women': Containment Culture and the Queer Astronomical Lyric," expands the first chapter's insights about the lyricization of the space age into a consideration of this moment's impacts on mid-century lyric. I turn to the careers of two seemingly unalike poets—Elizabeth Bishop and Adrienne Rich—to reveal how space age politics and aesthetics inflected the development of a mid-century queer lyric "I."

Chapter 3 moves to the moon. Lyric poetry, *the* genre of the moon, continues to play a major public relations role in elevating and commemorating lunar missions, even as many American poets have expressed dismay at NASA's demystification of the poetic moon. The chapter centers on the poetry of Agha Shahid Ali, who blends Urdu and Anglo-American poetic forms to transform the lunar conquest not into a celebration but, rather, into an elegy for global losses under Anglo-American empire. Chapter 4 continues with poets who encounter American imperialism from the perspective of other national situations and who turn to the metaphor of gravity to articulate cosmopolitan poetics against the apparent solipsism of Anglo-American lyric. Chapter 5 ends in deep space, with Tracy K. Smith's astronomical elegies for her father, a NASA engineer, through which she develops strategies of opacity to navigate gendered and racialized dimensions of the lyric "I." Her work is emblematic of a larger phenomenon in modern and contemporary poetry, especially African American poetry, of engaging scientific metaphors to imbue lyric speakers with an opacity that resists collective representation and assimilation. The Coda considers the practice of sending poems into space as an example of what I term "planetary elegy," a phenomenon that at once immortalizes the species as part of a nationalist fantasy *and* mourns our impending demise.

Lyric Poetry and Space Exploration from Einstein to the Present focuses on poets who work, often uneasily, within lyric's universal and estranging realms. In the contradictions that drive the postwar American lyric, poets confront the political limits and possibilities of lyric forms and figures. Lyric tools can equally well exalt or undermine American imperialism as it reaches for the stars. Even as it can drive rhetoric of American exceptionalism and universality, the lyric offers modes of relating to the self and others that emphasize not epistemological mastery but, rather, the limits of knowledge. The ensuing chapters follow lyric forms and figures across the cosmic scale, from the immense to the intimate.

1
The Lyrical Planet
Global Aesthetics and Planetary Ethics

At the close of 1968, planet Earth became the universe, and America. While in lunar orbit on Christmas Eve, the crew of Apollo 8 emerged from the dark side of the moon to see Earth shimmering against the wasteland of the lunar surface and starless space. "Oh my God! Look at that picture over there!" said astronaut Bill Anders, framing this view of Earth as a work of art.[1] Anders then took NASA image AS08-14-2383, nicknamed *Earthrise*. This photograph of the blue Earth rising above the monochromatic moon is credited with inaugurating a new planetary consciousness by showing Earth as a finite, vulnerable astronomical object—a planet among many, perhaps, but the only viable one for humans (see Figure 1.1).

Earthrise and similar photographs suggest a neo-Copernican cosmology. While other NASA photographs of Earth taken from space have also become icons of the Anthropocene, *Earthrise* in particular reversed the trajectory of astrofuturism and inaugurated the "age of ecology" by suggesting that Earth is the human species' only possible cosmic home.[2] *Earthrise* symbolized a new cultural imagination of the planet; it was published at the beginning of a four-year span that also saw Buckminster Fuller's *Operating Manual for Spaceship Earth* (1969), the first Earth Day (1970), and James Lovelock's Gaia hypothesis that proposed Earth to be a single organism (1972). *Earthrise* and its cousin *Blue Marble* (Apollo 17, 1972; see Figure 1.2), widely cited as history's most reproduced photograph, are familiar images of the Anthropocene, the proposed geological epoch in which humans drive planetary change through processes of turbo-capitalism. Ursula K. Heise notes the irony of these photographs' ties to ecology, emphasizing "their technological—indeed, to some

---

[1] *American Experience*, season 31, episode 4, "Chasing the Moon, Part Two," written and directed by Robert Stone, featuring Buzz Aldrin, George Alexander, and Bill Anders, aired 10 July 2019, https://www.pbs.org/video/chasing-the-moon-part-2-7s7mhp/.

[2] For a detailed history of *Earthrise*, see Robert Poole, *Earthrise: How Man First Saw the Earth* (New Haven: Yale University Press, 2010). Poole credits *Earthrise* with inaugurating the age of ecology on p. 9.

Figure 1.1 "Earthrise" (1968), NASA image AS08-14-2383. Original found at the NASA gallery: https://www.nasa.gov/multimedia/imagegallery/image_feature_1249.html.

extent, military—origin."[3] In short, planet Earth lives new and various lives in the wake of space exploration. Since the 1960s, it has been at once an aesthetic object, an icon of environmentalism, and an American imperial artifact.

It has also been a lyric poem. This chapter argues that planet Earth became a new kind of object in the late 1960s—specifically, a lyrical one. From 1967 to 1969, *National Geographic*, the *New York Times*, *Life*, and the *Times* printed

---

[3] Ursula K. Heise, *Sense of Place and Sense of Planet: The Environmental Imagination of the Global* (Oxford: Oxford University Press, 2008), 20. She interrupts her own critique of the photographs' nationalism to describe them lyrically: "Set against a black background like a precious jewel in a case of velvet, the planet here appears as single entity, united, limited, and delicately beautiful." For the most extensive critique of these photographs' nationalist and imperialist dimensions, see Tariq Jazeel, "Spatializing Difference Beyond Cosmopolitanism: Rethinking Planetary Futures," *Theory, Culture & Society* 28:5 (September 2011): 75–97. For a feminist critique of the image's militant imperialism, see Donna Haraway, *Modest_Witness@Second_Millennium.FemaleMan©_Meets_ OncoMouse: Feminism and Technoscience* (New York: Routledge, 1997): 174–8.

Figure 1.2 "Blue Marble" (1972), NASA image AS17-148-22727. Original found at the NASA gallery: https://earthobservatory.nasa.gov/images/1133/the-blue-marble-from-apollo-17.

the first photographs of planet Earth accompanied by lines of poetry, in effect training the public to read Earth lyrically. Lyric—as a genre, a mode, and a reading practice—also features in scholarly, civic, and artistic accounts of the post-1968 Earth. My contention that lyric figures and forms communicated a new planet Earth to the public might sound absurd. Particularly in the American context, the lyric's reputation as an aggressively apolitical, navel-gazing genre of individual expression makes it seem the *least* likely form to articulate new imperial fantasies, collective ideals, and planetary responsibilities. Besides, as the story goes, people don't read poetry, making the lyric an improbable tool for spreading nationalist sentiment or bolstering collective action. But as I will show, the public has been reading Earth lyrically since the 1960s. Examining a vast archive of historiography, popular magazines, astronaut accounts, and poems that lyricize these images, I demonstrate how and why lyric shaped a new cultural imagination of planet Earth that vacillates between global normativity and planetary strangeness.

Wallace Stevens anticipated the lyrical quality of planet Earth images at the dawn of the space age, over a decade before they hit the public with revelatory force. In the self-elegy "The Planet on the Table," Stevens's collected poems become a planet displayed for their creator's admiration. The poem and the planet share the eerie quality of a still life, suspending narrative time and space even as shadows in the margins and overripe images and language suggest the passing of time. Stevens rejects the ahistorical and immortalizing impulse of the lyric: "It was not important that they survive," he writes, so long as the poems evince "some lineament or character.../ Of the planet of which they were part."[4] Rather than becoming its own autonomous world, Stevens's planet on a table registers the world around it. Bonnie Costello argues for the political usefulness of still life in *Planets on Tables: Poetry, Still Life, and the Turning World*. "In times of public disturbance," she writes, still life can "provide a medium by which individuals might encounter historical realities that were otherwise too distant, too vast, too mediated, too dangerous, or too impersonal to feel and comprehend."[5] The realities of 1968 fit the bill. While pandemic influenza killed over one million people worldwide, the fallout of Anglo-American imperialism ravaged the globe it had made in its image: the U.S. escalated aggressions in Vietnam as British colonial practices gave way to the atrocities of the Biafran Civil War. Within the U.S. there were strikes, protests, and demonstrations. The assassinations of Martin Luther King, Jr., and Robert F. Kennedy, wrenching in their moment, also recalled the 1963 assassination of John F. Kennedy, who had launched the space race in a tone of youthful optimism. Then came the planet, dropped into that world. It, too, was a new thing, "too distant, too vast...too dangerous...too impersonal" to apprehend.

---

[4] Wallace Stevens, "The Planet on the Table," in *The Palm at the End of the Mind: Selected Poems and a Play*, ed. Holly Stevens (New York: Vintage, 1967), 386. The poem was first published in 1954, a year before the poet's death.

[5] Bonnie Costello, *Planets on Tables: Poetry, Still Life, and the Turning World* (Ithaca, NY: Cornell University Press, 2008), viii.

Removed from the flux of history and politics, suspended in a contextless void, Earth pictured in space invited what Virginia Jackson calls "lyric reading"—a method that shuts out context to create the illusion of an autonomous work of art.[6] This is the lyric of John Stuart Mill's famous description, the "overheard" utterance of an expressive speaker or "feeling confessing itself to itself, in moments of solitude."[7] While Jackson and others apply lyric reading to understanding poetry—specifically, to understanding the historical processes that led distinct types of poems to be "lyricized"—I suggest that any number of texts can be, and frequently are, subject to lyric reading. In calling *Earthrise* and *Blue Marble* lyric objects, I mean to describe two separate phenomena. First is the way in which Apollo-era institutions opportunistically lyricized these photographs to elide grim material realities, conveying that American democracy had harmonized rather than ravaged the world. Second, I contend that these photographs have formal properties that lent them to lyricization in the first place, and, counterintuitively, some of their endemic lyricism undercuts the assimilative logic of globalization. This claim relies on a stranger account of lyric reading than allowed by Jackson's historical materialist poetics. By lyric reading, I do not just mean reducing a text to an autonomous work in an interpretive void of space. To read lyrically is also to foreground the very strange behaviors of the lyric, in fact to offer its figures of estrangement as tools for reading the world.

This opening chapter does not aim to offer a theory of the lyric in a globalized world, nor a theory of global or transnational literature, but rather a cultural history of how and why the public came to read planet Earth as a lyric object. Approaching these photographs lyrically reveals "planet Earth" to be at once an American Cold War artifact and a strange, unassimilable otherworld. This chapter brings to light the ideological work lyric was called upon to perform as part of a messianic, ostensibly peaceful, conquest of space on behalf of humankind. It will also examine how lyric planetarity undermines an assimilative collective vision of the human. As I explore in ensuing chapters, this tug between global aesthetics and planetary ethics has had a profound impact on the development of twentieth- and twenty-first-century lyric poetry in the shadow of Anglo-American imperialism.

---

[6] Virginia Jackson, *Dickinson's Misery: A Theory of Lyric Reading* (Princeton, NJ: Princeton University Press, 2005), 100.

[7] John Stuart Mill, *Essays on Poetry*, ed. F. Parvin Sharpless (Columbia, SC: University of South Carolina Press, 1976), 12.

1.1 The Planet and the Nation

Planet Earth rarely features directly in accounts of the nation. As an astronomical and geological body and a metaphor for relationality, the planet has been the central figure of the "planetary turn" across the humanities. In their attempt to define this emerging field, Amy J. Elias and Christian Moraru "define 'planet' and 'planetarity' as a noun and an attribute signifying and qualifying, respectively, *a multicentric and pluralizing, 'actually existing' worldly structure of relatedness critically keyed to non-totalist, non-homogenizing, and anti-hegemonic operations typically and polemically subtended by an eco-logic*" (original emphasis).[8] By focusing on ontology, alterity, and ecology, Elias and Moraru attempt to distinguish planetarity from neighboring ideas of globalization and cosmopolitanism. Yet a driving force of their ambitious anthology of planetary essays is neither the globe nor cosmopolitanism but, rather, America: the editors periodize the planetary as a post-Cold War phenomenon, and the volume's disproportionate focus on American literature reproduces what Paul Giles describes as an "abstract universalism of U.S. critical discourse on the planet."[9] Elizabeth DeLoughrey's work helps to make sense of the specter of America in planetary studies. She argues that nationalist, planetary, and imperial discourses tied to the Cold War are inseparable: the very Cold War military technologies that produced *Earthrise* and *Blue Marble* led to knowledge of climate change and theorization of the Anthropocene, and indeed the U.S. military leads the world as "the largest institutional consumer and producer of fossil fuels and carbon emissions."[10] In short, DeLoughrey notes, views of Earth's biosphere come to us through the very military technologies contributing disproportionally to climate change.[11] Moreover, environmentalist rhetoric often melds Christian and American redemptive narratives through eco-nationalism: take Ronald Reagan's "America the beautiful" or the messianic call to "save the planet." *Earthrise* in some ways visualizes Earth as a national park protected by America, positioning America as the savior of the world.

---

[8] See Amy J. Elias and Christian Moraru, *The Planetary Turn: Relationality and Geoaesthetics in the Twenty-First Century* (Evanston, IL: Northwestern University Press, 2015), xxiii.

[9] Elias and Moraru, *The Planetary Turn*, xxiv; Paul Giles, "Writing for the Planet: Contemporary Australian Fiction," in *The Planetary Turn*, 143–60, here 147.

[10] Elizabeth DeLoughrey, *Allegories of the Anthropocene* (Durham, NC: Duke University Press, 2019), 69.

[11] Elizabeth DeLoughrey, "Satellite Planetarity and the Ends of the Earth," *Public Culture* 26:2 (March 2014): 257–80. DeLoughrey writes, "our evidence for and understanding of the Anthropocene has been produced by the very military technologies that brought us the Cold War" (274).

Some planetary work has inadvertently bolstered national power by naturalizing it through scientific metaphors. In *Through Other Continents*, Wai Chee Dimock proposes the metaphor of "deep time"—geology's and astronomy's unthinkably vast scales that eclipse human time—to conceptualize "the input channels, kinship networks, routes of transit, and forms of attachment...[that] thread American texts into the topical events of other cultures, while also threading the long durations of those other cultures into the short chronology of the United States." Dimock invokes deep time to argue that "American literature emerges with a much longer history than one might think."[12] However, as Tim Wientzen points out, the human cultural time containing Dimock's texts accounts for only 1 percent of human existence, 0.0000004 percent of the planet's life as a geological entity,[13] or about 0.0000001 percent of the life of the universe. Dimock's account of planetarity does not attend to the planet's otherness, often experienced by confronting the impossible-to-think scales that Dimock regularizes into national and global temporalities. Deep time metaphors risk extending American power boundlessly into the inconceivably deep past and the deep future.

The most famous and influential account of planetarity arose in reaction to American universities' promotion of global education in the 1990s. Gayatri Spivak's foundational work in planetarity challenges an Anglo-American vision of a human collective on a homogeneous sphere by "propos[ing] the planet to overwrite the globe." Outwardly, Spivak's theory concerns the relationship between the global and the planetary without recourse to the national: "Globalization is the imposition of the same system of exchange everywhere.... The globe is on our computers. No one lives there. It allows us to think that we can aim to control it. The planet is in the species of alterity, belonging to another system; and yet we inhabit it, on loan."[14] To recognize our relation with the planet as one of alterity requires unmooring the Western subject from its position as a global agent. For this, Spivak argues, we need an extraterrestrial framework: "Planetary imaginings locate the imperative in a galactic and para-galactic alterity—so to speak!—that cannot be reasoned into the self-interest that extends as far as recognizing the self-consolidating

---

[12] Wai Chee Dimock, *Through Other Continents: American Literature Across Deep Time* (Princeton: Princeton University Press, 2006), 3–4.

[13] Timothy Wientzen, "Not a Globe but a Planet: Modernism and the Epoch of Modernity," *Modernism/Modernity* 2:4 (2018), https://doi.org/10.26597/mod.0039.

[14] Gayatri Chakravorty Spivak, "Planetarity," in *Death of a Discipline* (New York: Columbia University Press, 2003), 72.

other as the self's mere negation."[15] That is, in extraterrestrial space, the "I" finds nothing to master against which to define itself; it encounters only alterity. This part of Spivak's thinking emerged as a corrective to American universities' flattening "global studies" curriculum that erased specificities of so-called "minority" languages and cultures by centering an implicitly white American "I" who engages in self-actualizing encounters with cultural others.[16] Her focus on planetarity, then, is a corrective not only to globalization but also, implicitly, to U.S. cultural imperialism that treats the globe as its domain.

Increasingly, critics have attended to the enmeshment of the national, the global, and the planetary. Challenging Spivak's distinction between the globe and the planet, Dipesh Chakrabarty emphasizes the entanglement of planetary thinking and turbo-capitalism, driven in the postwar era by the U.S.: we are able to imagine the planet, Chakrabarty points out, only because of how these processes have brought the species to this technological point. He argues that the rise of Earth System Science (developed during the space race) "is not an end to the project of capitalist globalization but the arrival of a point in history where the global discloses to humans the domain of the planetary."[17] Global and planetary processes and language are so ensnared that, "[f]or all their differences, thinking globally and thinking in a planetary mode are not either/or questions for humans."[18] In the era in which the U.S. has positioned itself as the global norm, largely through the activities and rhetoric of space exploration, to think of the planet is also to think of the nation.

The development of the post-1968 lyric records how the global, the planetary, and the national inflect and unsettle one another. Lyric poetry has now made its way into global and planetary studies despite the long dominance of the novel and the epic in these fields. Jahan Ramazani paved the way with *A Transnational Poetics*, where he argues, drawing from David Harvey, "By virtue of its extraordinary compression, poetry readily evinces the 'time-space compression' of globalization."[19] Formal properties of much lyric poetry—including compression and patterns of stanza, line, and sound—resemble

---

[15] Gayatri Chakravorty Spivak, "The Imperative to Re-imagine the Planet" (1999), in *An Aesthetic Education in the Era of Globalization* (Cambridge, MA: Harvard University Press, 2012), 340.

[16] Spivak discusses this context in *Death of a Discipline*, 72.

[17] Dipesh Chakrabarty, "The Planet: An Emergent Humanist Category," *Critical Inquiry* 46:1 (Autumn 2019): 1–31, here 17. For an account of the impact of the space race on the development of planetary science, see Lisa Messeri, *Placing Outer Space* (Durham, NC: Duke University Press, 2016), 5–7.

[18] Chakrabarty, "The Planet," 23.

[19] Jahan Ramazani, *A Transnational Poetics* (Chicago, IL: University of Chicago Press, 2009), 16.

frictionless global space. Formal poetry in particular evinces what Ramazani, following Spivak's language, calls "the abstract geometry of the global," the neat flows and circuits between one part of a poem and the next that create the sense of a harmonious aesthetic object, recalling the cosmic divinity with which the Pythagoreans imbued geometry.[20] Yet lyric form and figures also lend themselves to planetary thinking. Spivak's planetarity is, in fact, a poetic theory. Spivak finds the planet, rather than the globe, to be a poetic object, a figure "impossible" to apprehend through Western scientific modes of knowledge. In her account, thinking about the planet requires a mode of *teleiopoiesis* (adopted from Derrida) rather than *istoria* (learning through knowledge). In conceiving of the planet as a hard-to-read poetic figure, Spivak turns to Heidegger's ideas of poetry, developed in reaction to what he called the "age of the world picture" visualized by Earth photography from space. "I... was frightened when I saw pictures coming from the moon to the earth. We don't need any atom bomb. The uprooting of man has already taken place," Heidegger complained.[21] These photographs exemplified for him the culmination of a project of "planetary imperialism," evincing "total... technological rule over the earth" that cut humans off from experiencing dwelling.[22] Heidegger proposes that poetry, or *poiesis*, meaning *bringing into being* or *making*, offers an opportunity for genuine thought outside of the rote procedures of the scientific method that become all-consuming in the technoscientific landscape of modern life.[23] Following Heidegger, Spivak proposes that poetic language resists familiar Western modes of epistemological mastery associated with the global.

While Ramazani focuses on poetry's play with scale in the context of the global, scale is also a core concern of planetary thinking. Lyric poetry grapples not only with the immense though calculable scales of the global but also with planetary scales unmasterable, even unimaginable, by national powers. While Dimock argues that the ancient, "scale-rich, scale-variable" epic is particularly well suited to model the scales of the planetary, lyric forms and

---

[20] Ramazani, *A Transnational Poetics*, 17.

[21] Martin Heidegger, "'Only a God Can Save Us': *Der Spiegel*'s Interview with Martin Heidegger (1966)," in *The Heidegger Controversy: A Critical Reader*, ed. Richard Wolin (Cambridge, MA: MIT Press, 1992), 105–6. For more on Heidegger's responses to planet Earth photography, see Benjamin Lazier, "Earthrise; or, The Globalization of the World Picture," *American Historical Review* 116:3 (June 2011): 602–30, and Kelly Oliver, "The Earth's Refusal: Heidegger," in *Earth and World: Philosophy After the Apollo Missions* (New York: Columbia University Press, 2015), 111–62.

[22] Martin Heidegger, "The Age of the World Picture," in *The Question Concerning Technology and Other Essays*, trans. William Lovitt (New York: Harper, 1977), 152.

[23] Heidegger, "...Poetically Man Dwells...," in *Poetry, Language, Thought*, trans. Albert Hofstadter (New York, HarperCollins, 2001), 219, 224–5.

figures capture how disorienting it is to think in terms of scales that at once exceed and include the human.[24] Lyric poetry offers tools to address what Chakrabarty describes as the great cognitive challenge of the Anthropocene, "to think of human agency over multiple and incommensurable scales at once," as "humans are now part of the natural history of the planet."[25]

While globalization, like planetarity, contends with vast scales, the global view suggests that the subject can master these scales rather than be undone by them. In contrast, the figure of the planet so radically exceeds human comprehension that it approaches the sublime. Timothy Morton brings planetary thinking and the sublime together, even suggesting that the planet is a sublime object. Specifically, the planet is a "hyperobject," something "massively distributed in time and space relative to humans" that exceeds our comprehension and undoes our agential position over the planet.[26] Morton excavates Longinus's original formulation of the sublime to argue for a theory of the planet that decenters the human and emphasizes the strangeness of all objects to one another. In a sublime tradition reaching back to Longinus's readings of Sappho's poetry (as the Introduction explored), the lyric, for all its association with global normativity, carries within it a countervailing tradition of the alien and the strange that is the domain of the planetary. This divided but entangled tradition—global assimilation and planetary estrangement—has shaped ensuing poetic responses to extraterrestrial space and, indeed, to the politics and aesthetics of contemporary lyric poetry.

1.2 Planetary Apostrophe

Rendered in brilliant visual transparency and occluded in indecipherable meanings, the planet invites, even demands, lyrical figuration to contend with its otherworldliness. Apostrophe is the most obvious and most used of these figures. A stunning number of poems in the Cold War and after conjure the planet through the "O" of apostrophic address, an updated *Earthrise*-era version of the "O moon" trope enjoyed by English poets since the sixteenth century.

[24] Wai Chee Dimock, "Gilgamesh's Planetary Turns," in Elias and Moraru, *The Planetary Turn*, 131.
[25] Dipesh Chakrabarty, "Postcolonial Studies and the Challenge of Climate Change," *New Literary History* 43:1 (Winter 2012): 1, 10.
[26] See Timothy Morton, *Hyperobjects: Philosophy and Ecology after the End of the World* (Minneapolis: University of Minnesota Press, 2013), 3. Morton first develops the concept in Timothy Morton, *The Ecological Thought* (Cambridge, MA: Harvard University Press, 2010), 130–5. For the relationship of the sublime to hyperobjects, see Timothy Morton, "Sublime Objects," in *Speculations II* (Santa Barbara, CA: punctum books, 2011), 207–27.

When Sir Philip Sidney's speaker cries, "With how sad steps, O Moon, thou climb'st the skies!" he of course does not expect the moon to hear, and that's not really what he wants to tell it anyway. "O little Earth," Adrienne Rich wrote presciently in 1953 from her lyric speaker's position on the moon, shrinking Earth to the size of the graphic apostrophe while extending the reach of an ancient lyric figure across extraterrestrial space.[27] In photographs like *Blue Marble* that lack the contrasting point of the moon, the planet seems to address its readers directly, gaping in its spherical "O," at once indifferent to us and requesting a response. Like the "messages" of many lyric poems, the planet's is at once crystal clear and indiscernible—the planet gazes at us with something not unlike W. H. Auden's definition of poetry as "the clear expression of mixed feelings."[28]

Besides the serendipity of the O shape, why apostrophe? In crying out to a thing that cannot respond, apostrophe is a figure of yearning and unattainability: it posits an impossible relationship between two things. *Earthrise*, with its two coordinates of the Earth and the moon, visualizes the structure of lyric address, an "I" calling out to, or overheard by, an unresponsive "thou" suspended in a void. Viewers often feel that *Earthrise* and *Blue Marble* have something to tell them, if they could only speak back: perhaps a revelation of humanity's place in the cosmos, an ecological warning, or a call for collective action. The planet, seen inconceivably from beyond the sphere that created and shelters human life, stirs nostalgia in its very unresponsiveness. Seen from beyond, the Earth is both our home and an alien world.

Apostrophe is a fundamentally relational figure, an I calling to a thou, an Earth to a moon, a human to a planet. In its way of "foregrounding the lyric as act of address," as Jonathan Culler puts it, apostrophe orients the lyric outward as much as inward, suggesting that the lyric, for all its purchase on solitude, is particularly adept at exploring relations with others and otherness.[29] In *Earthrise*, the "I" and the "thou"—moon and Earth, planet and human—become enmeshed through address. Since Barbara Johnson's 1986 essay "Apostrophe, Animation, and Abortion," critics have highlighted how apostrophe breaks down coherent subject–object positions. In animating others who would not conventionally "count as a person," Johnson argues,

---

[27] Adrienne Rich, "The Explorers," in *Collected Poems: 1950–2012* (New York: W.W. Norton, 2016), 63.

[28] W. H. Auden, "New Year Letter," in *Collected Longer Poems of W. H. Auden* (New York: Random House, 1965), 79.

[29] Jonathan Culler, *Theory of the Lyric* (Cambridge, MA: Harvard University Press, 2015), 213.

apostrophe structures debates about personhood, power, and justice.[30] Margaret Ronda applies Johnson's insights about apostrophe and ethics to human–nonhuman relations. Drawing on Bruno Latour, Ronda describes how the figures of apostrophe and prosopopoeia reveal "previously unfathomed dimensions of the human subject" as humans in the Anthropocene confront "the disconcerting sense of being 'morphed' by one's own creation." Ronda demonstrates how figures of address—key features of this "anthropogenic poetics"—"occur as extravagant or ironized calls, admissions of guilt and shame, or refusals or inabilities to address another."[31] As culture at large and individual poems have turned to face the planet since 1968, they have done so with all of this ambivalence Ronda identifies, as well as with a mood of what Paul Gilroy calls "hopeful despair" that he attaches to the Apollo images.[32]

Though this book is the first study to recognize the far-reaching cultural implications of the phenomenon, I am not the first to notice that planet Earth suspended in space graphically resembles apostrophe. Ramazani, Sumita Chakraborty, and Joshua Schuster—in addition to the myriad poets explored in this book—each examine an effect that I refer to as "planetary apostrophe." Ramazani identifies a subgenre of "planetary poems" since the 1960s that "peer down on the Earth from beyond its surface and figure the world as 'O,' or maybe 'zero.'" Poems like Seamus Heaney's "Alphabets" (which I explore in Chapter 4), featuring an astronaut gazing at Earth from lunar distance, recall the O "of literary artifice, the apostrophic 'O' of poetry, as if in an echo of the roundness of a wonder-inspiring world."[33] Chakraborty links Heaney's apostrophe not to the globe but rather to Spivak's uncanny planet, arguing that apostrophe itself is an uncanny figure that also brings lyric itself into the domain of the planetary.[34] Moving planetary apostrophe into deep space, Schuster brings together ecopoetics and the extraterrestrial in his neologism "exopoetics," which describes "the interface of poetry with SETI [the Search for Extra-Terrestrial Intelligence] and SETI with poetry." Arguing that SETI and poetics can inform each other's modes of communication, he suggests

---

[30] Barbara Johnson, "Apostrophe, Animation, Abortion," *Diacritics* 16:1 (Spring 1986): 28–47.
[31] Margaret Ronda, *Remainders* (Stanford, CA: Stanford University Press, 2018), 113–14.
[32] Paul Gilroy, *Postcolonial Melancholia* (New York: Columbia University Press, 2004), 72. This view of Earth, Gilroy writes, "supports an appreciation of nature as a common condition of our imperiled existence, resistant to commodification and...deeply incompatible with the institution of private property that made land into a commodity and legitimized chattel slavery."
[33] Ramazani, *A Transnational Poetics*, 16–17.
[34] Sumita Chakraborty, "Of New Calligraphy: Seamus Heaney, Planetarity, and Lyric's Uncanny Space-Walk," *Cultural Critique* 104:14 (Summer 2019): 101–34.

that many poems, regardless of subject matter, might be read as "exopoetic" in how they "reflect in their form and content issues of durability, readability across cultures, and the mechanics and emotional investments in casting out expressions to an unknown fate."[35] Tracy K. Smith captures the poignance of SETI-like lyric address in a metapoetic line from "My God, It's Full of Stars": "*This message going out to all of space*," she writes, allowing the message to trail off in a line that sends it, probably futilely, into the unknown cosmic "it" of the poem's title.[36] Astronomical lyric delves into unknowable realms of both inner and outer alienness, often as a corrective to narratives of global domination that emphasize mastering the planet and all the peoples on it.

A range of poems evoking planetary apostrophe explore how and how *not* to relate to the planet. Allen Ginsberg uses planetary apostrophe to highlight American imperialism as a driving force of the space age, a period during which the FBI watched him for un-American behavior. In his anti-moon landing poem from *The Fall of America*, he writes of the dystopian imperialist views of planet Earth: "No Science Fiction expected this Globe-Eye Consciousness."[37] He dismisses the feat by suggesting that it's hardly exceptional or new to "see" Earth: "Saw the Earth in dream age 37, half cloud-wrapped, from a balcony in outer-space"[38]—a "vision" he had five years before *Earthrise*. For Ginsberg, to militarize moon was an affront against Earth: "A new moon looks down on our sweet sick planet" and "They fucked up the planet!" and "Oh awful man! What have we made the world! Oh man / Capitalist exploiter of Mother Planet!"[39] Instead of "O God," or the "O Moon" of poetic devotion, we get a sigh—"Oh man"—at the godlike entity that has ruined Earth.

Other poets use planetary apostrophe to record Anglo-American imperial expansion as it diminishes the globe. The space program aimed to communicate American ideals of "freedom" and "unity" to the emerging postcolonial world to entice new nations to join its democratic mission. This Cold War context had a profound impact on *Earthrise* poetry. In fact, the most prolific *Earthrise* poet is the anglophone Indian writer Satya Dev Jaggi. In the U.S., the best-known planet Earth poet is probably A. R. Ammons, whose

---

[35] Joshua Schuster, "Another Poetry Is Possible: Will Alexander, Planetary Futures, and Exopoetics," *Resilience: A Journal of the Environmental Humanities* 4:2–3, Environmental Futurity (Spring–Fall 2017), 147–65, here 154. Schuster frames the symbiosis between "poetry and planetary thinking" through the way in which some poems reflect "on how *form*, *content*, and *contact* are interconnected."
[36] Tracy K. Smith, *Life on Mars* (Minneapolis, MN: Graywolf Press, 2011), 7.
[37] Allen Ginsberg, *The Fall of America: Poems of These States* (San Francisco, CA: City Lights, 1974), 127–9.
[38] Ginsberg, *The Fall of America*, 128. [39] Ginsberg, *The Fall of America*, 80, 167.

volume-spanning long poem *Sphere* (1974) contemplates the scales and motions of Earth using shifting vantage points that range from gas stations to galaxies.[40] But Ammons has nothing on Jaggi's commitment to these images: Jaggi produced an incredible four volumes of *Earthrise* poems from 1969 to 1970. Writing as India was developing its own space program while the U.S. and USSR. battled for control of the subcontinent, Jaggi at once praises Apollo astronauts for their scientific-aesthetic vision and compares them unfavorably to "Columbuses of Space" whose explorations have reduced earth to "[a] compact sphere against those sunless stars." Jaggi evokes the shrunken globe through planetary apostrophe: "O you have shrunk the proportions of / My longing heart."[41] Rather than praising scientific expansion, the lines lament the way in which space exploration diminishes both the human and the world. Jaggi's ambivalence about the Apollo missions—they are at once aesthetic achievements and imperial adventures that literally and imaginatively shrink the globe—is a hallmark of postcolonial poetic responses to these images, which I explore further in Chapter 3 through the work of Kashmiri American poet Agha Shahid Ali.

A number of Anglo-American and anglophone poems are metapoetically invested in learning to read the planet through lyric figures, foregrounding the act of lyric reading through the planetary apostrophe. These include Auden's "Prologue at Sixty" (1967) as well as a range of later poems I consider in ensuing chapters, including Bishop's "In the Waiting Room" (1971), Derek Walcott's "The Fortunate Traveller" (1981), and above all Heaney's "Alphabets" (1987).[42] All of these poems metapoetically explore the act of reading lyric poetry to instruct their readers in how to relate to others on a shared planet. Auden's "Prologue at Sixty," an early self-elegy published just as photographs of Earth from space were beginning to appear in the popular press, features the emigrant poet reading the newspaper and wondering if he ought to classify himself now as an American or on the smaller scale of a New Yorker. The poem moves from local identification to the biggest conceivable scale of belonging, that of the entire Earth. Auden records the poem's own failure to

---

[40] For a reading of *Sphere* in the context of the *Earthrise* moment, see Kevin McGuirk, "A. R. Ammons and the Whole Earth," *Cultural Critique* 37 (Autumn 1997): 131–58.

[41] Satya Dev Jaggi, "After the Moon Mission of Apollo 8," in *The earthrise and other poems* ([Delhi:] Falcon Poetry Society, 1969), 20. His other Earthrise volumes are *One looks earthward again* ([Delhi:] Falcon Poetry Society [1970]); *The moon voyagers and other poems* ([Delhi:] Falcon Poetry Society [1970]); and *Our awkward earth* ([Delhi:] Falcon Poetry Society [1970]).

[42] Ramazani identifies these poems by Auden, Walcott, and Heaney as "planetary poems" that assume a "literally *cosmo*politan perspective" in figuring the Earth as O. See *A Transnational Poetics*, 16.

read Earth at this scale through the unmoved eyes of Laika, the dog launched into orbit on a USSR test flight and left to die of dehydration.[43] "Already a helpless orbited dog / has blinked at our sorry conceited O," Auden writes, the roundness of the dog's eye picked up in the figure of the planet it apprehends.[44] The apostrophe, stranded at the end of the line as Laika is stranded in space, fails to launch its communicative circuit between speaker and listener. The Earth from this perspective is unsummonable, unreadable, cut off from the possibility of ethical reasoning.

As in his other space age poems, Auden evokes his friend Hannah Arendt, who shared with Heidegger (her teacher) a sense that science was outscaling human cognition. Arendt begins *Human Condition* with the launch of Sputnik, which she considers "second in importance to no other, not even to the splitting of the atom."[45] Leaving the planet, she writes, is an affront against the human condition. Only on Earth can the species "move and breathe without effort and without artifice;" leaving means making life "artificial" and severing human ties to "nature."[46] Moreover, the technological and scientific feats necessary to leave the planet mean that humans can no longer comprehend our environments: she worries that "we, who are earth-bound creatures and have begun to act as though we were dwellers of the universe, will forever be unable to understand, that is, to think and speak about the things which nevertheless we are able to do."[47] The inability to discuss the science that we are nonetheless acting on undermines political participation and ethical debate, Arendt argues. Auden shares Arendt's concerns and captures them throughout his space age Horatian odes, a scale-intense form of high artifice that uses alcaics (quatrains with the precise syllabic count of 11-10-9-10) to capture ethical and cognitive problems of scale. His "Ode to Terminus" begins with a pun at the level of meter and line length, the scales of his lines expanding and contracting:

> The High Priests of telescopes and cyclotrons
> keep making pronouncements about happenings
>   on scales too gigantic or dwarfish
>   to be noticed by our native senses.[48]

---

[43] Alice George, "The Sad, Sad Story of Laika, the Space Dog, and Her One-Way Trip into Orbit," *Smithsonian Magazine*, 11 Apr. 2018, https://www.smithsonianmag.com/smithsonian-institution/sad-story-laika-space-dog-and-her-one-way-trip-orbit-1-180968728/.

[44] W. H. Auden, *Collected Poems,* Centennial Edition, ed. Edward Mendelson (New York: Random House, 2007), 832.

[45] Hannah Arendt, *The Human Condition*, 2nd ed. (Chicago, IL: University of Chicago Press, 2013), 1.

[46] Arendt, *The Human Condition*, 2.   [47] Arendt, *The Human Condition*, 3.

[48] Auden, *Collected Poems*, 809.

Auden plays a similar game with scale and form in his early moon poem "This Lunar Beauty," where each of the three stanzas adds a line in a pun on "waxing poetic."

Related to his distrust of the scales of science that eclipse human experience, Auden's work abounds with views from airplanes and spacecraft taking in planet Earth at different scales. One would think, says Auden, that seeing Earth in full would break down boundaries: "From the height of 10,000 feet, the earth appears to the human eye as it appears to the eye of the camera; that is to say, all history is reduced to nature. This has the salutary effect of making historical evils, like national divisions and political hatreds, seem absurd."[49] But Auden qualifies this sentiment with the realization that zooming out creates feelings of disconnection from Earth that lead to terrible ethical decisions:

> Unfortunately, I cannot have this revelation without simultaneously having the illusion that here are no historical values either. From the same height I cannot distinguish between an outcrop of rock and a Gothic cathedral, or between a happy family playing in a backyard and a flock of sheep, so that I am unable to feel any difference between dropping a bomb upon one or the other.[50]

The bomb in this gentle, pastoral scene speaks to the military history and technology of the space race, reminding us that air and space technology, as well as the bird's-eye view in the visual and poetic arts, came out of aerial bombardment technology. This context is also behind Auden's planetary apostrophe in "Prologue at Sixty." The failed apostrophe in "Prologue at Sixty" renders planet Earth—a sorry, conceited place destroyed by an "anxious species" that sacrifices others for military ends masquerading as science—as cut off from human relations, unsummonable.

1.3 Lyrical Astronauts and Apollo's Eye

"Oh my God!" were Anders's first words to—or rather, at—the planet in the moments before taking his famous *Earthrise* photograph. Moved by the sight of the unattainable Earth 384,400 kilometers away, he cried out in the ritualistic utterance of prayer and lyric to a listener that cannot respond.[51] Anders's

---

[49] W. H. Auden, *Dyer's Hand and Other Essays* (New York: Vintage, 1989), 101.
[50] Auden, *Dyer's Hand and Other Essays*, 101.
[51] Helen Vendler, *Invisible Listeners: Lyric Intimacy in Herbert, Whitman, and Ashbery* (Princeton, NJ: Princeton University Press, 2005), 1.

excitement was amplified by his position—not just in seeing Earth from this perspective, but in being positioned to do so. Anders was never supposed to go to the moon. He was originally scheduled for Apollo 9, a high Earth orbit mission to test the command and lunar modules that should have taken place in early 1969. When numerous manufacturing errors derailed the low Earth orbit testing of the lunar module scheduled for Apollo 8 and threatened NASA's timeline for the moon landing, NASA rejiggered their mission plans to prevent yet another Soviet victory in the space race. Apollo 8 became a circumlunar mission for command module testing—including communications and trajectory calculations for reaching the moon and returning successfully to Earth—with Anders, Frank Borman, and James Lovell onboard.[52] The crew of Apollo 8 was also charged with documenting the lunar surface; as Anders had been slated to command the absent lunar module, the task of photography fell to him. When Anders saw "that picture" of the Earth, Borman replied, "Don't take that, it's not scheduled," needling Anders about his rigorous adherence to mission parameters while the photographer exchanged his black-and-white film for color.[53]

In its final issue of 1968, *Time* magazine's cover featured *Earthrise* with the caption "Dawn" to announce the rising of a new era of the *Pax Americana*—a message jarringly at odds with the national and global turmoil that led the *Washington Post* to call 1968 "the year that unraveled America."[54] While many believed that the Apollo 8 mission had "saved 1968," the distribution of *Earthrise* turned the photograph itself into a force of hope and redemption, a high note to end the year.[55] They communicate, in other words, the world-salvific promise of American democracy.

Behind this articulation of the American democratic promise is a long Western aesthetic-devotional-imperial tradition. In his influential book

---

[52] Melanie Whiting, "50 Years Ago: Considered Changes to Apollo 8," *NASA History*, 8 Aug. 2018, https://www.nasa.gov/feature/50-years-ago-considered-changes-to-apollo-8.

[53] "That was the most beautiful thing I'd ever seen," Anders told Andrew Chaikin in a 1987 interview. "Totally unanticipated. Because we were being trained to *go to* the Moon... It wasn't 'going to the Moon and looking back at the Earth.' I never even thought about that!... In lunar orbit, it occurred to me that, here we are, all the way up there at the Moon, and we're studying this thing, and it's really the Earth as seen from the Moon that's the most interesting aspect of this flight." See Andrew Chaikin, "Who Took the Legendary Earthrise Photo From Apollo 8?" *Smithsonian Magazine*, Jan. 2018, http://www.smithsonianmag.com/science-nature/who-took-legendary-earthrise-photo-apollo-8-180967505/.

[54] Marc Fisher, "1968: The year America unraveled," *Washington Post*, 29 May 2018, https://www.washingtonpost.com/graphics/2018/national/1968-history-major-events-in-pop-culture/#:~:text=1968%20was%20the%20year%20the%20center%20did%20not,yet%20also%20a%20time%20of%20passion%20and%20possibility.

[55] "50 years ago the Apollo 8 moonshot gave humans a new perspective of Earth," *Washington Post* online, 21 Dec. 2018, https://www.washingtonpost.com/lifestyle/kidspost/50-years-ago-the-apollo-8-moonshot-gave-humans-a-new-perspective-of-earth/2018/12/21/998c8db4-02f5-11e9-9122-82e98f91ee6f_story.html.

Apollo's Eye, geographer Denis Cosgrove traces the history of imagining the whole Earth in the West from ancient imperial dreams of rising above the terrestrial globe to the literal views of the planet provided by the Apollo missions. Cosgrove terms this vision of Earth from beyond it "Apollo's eye," after the god of the sun and poetry. The Apollonian gaze "seizes divine authority for itself, radiating power across the global surface from a sacred center, locating and projecting human authority imperially toward the ends of the earth."[56] This gaze is "individualized," Cosgrove explains, "a divine and mastering view from a single perspective" that "pulls diverse life on earth into a vision of unity." Cosgrove links the appeal of NASA photographs to ethical-aesthetic values of harmony in the West in which poetry, as a form linked to divination in ancient Greece, held a special place.[57] The Apollonian gaze positions an individual's vantage point as a universal truth in ways that resonate with postwar critiques of lyric subjectivity. Vitriol for the solipsistic, imperial lyric is especially leveled at formal poetry, as in Marjorie Perloff's complaint about the "transcendental ego" that arises from "the formation of a coherent or consistent lyrical voice."[58] While Cosgrove does not draw out the association between lyric poetry and the Apollonian gaze, *Apollo's Eye* opens with two stanzas of Robert Sabatier's poem *Icare*, and his opening chapter is named, tellingly, "The Imperial and Poetic Globe."[59] The multiple roles of the god Apollo are further suggestive of the connection between the Apollonian gaze and lyric poetry. As poet James Dickey put it in a program on the Apollo 11 mission produced by the American Broadcasting Company, "It seems a strange and wonderful thing to me that the spaceship that brought them to the moon should be named Apollo, who was also the Greek god of poetry."[60] Lyric itself is often visualized by images of Apollo with his lyre. Dickey associates lyric poetry with the Apollonian gaze to convey a nationalist message, unconsciously linking monologic lyric subjectivity to the universalized individual perspective of Apollo's eye that takes in and harmonizes the world below in a singular vision. The Apollonian gaze largely drives the imperial and lyrical power of Earth photographs.

According to Cosgrove, actually *seeing* Earth complicates the Apollonian gaze: "Associations historically attached to seeing the globe remain potent,

---

[56] Denis Cosgrove, *Apollo's Eye: A Cartographic Genealogy of the Earth in the Western Imagination* (Baltimore, MD: Johns Hopkins University Press, 2001), 256–65, xi.

[57] Cosgrove, *Apollo's Eye*, 29.

[58] Marjorie Perloff, *Poetic License: Essays on Modernist and Postmodernist Lyric* (Evanston, IL: Northwestern University Press, 1990), 12.

[59] Cosgrove, *Apollo's Eye*, 1.

[60] James Dickey, "Apollo 11: As it Happened – Poet James Dickey, 'The Moon Ground,'" *American Broadcasting Company*, 1969, https://www.youtube.com/watch?v=zaSGs8DQ_PQ.

but physically viewing the earth has generated the anxious paradox of a humanity at once isolated on a fragile solar satellite and lacking any special distinction from other life on that sphere."[61] To the universalizing "one-world" discourse that emphasizes the "equality of all locations networked across frictionless space," the Apollo photographs add a "whole-earth" discourse that emphasizes dwelling, ecological connections, and planetary stewardship.[62] Cosgrove notes that NASA Earth images became so popular in part due to their ties to the ancient Greek cosmos: "The image's geographical, compositional, and tonal qualities give it unusually strong imaginative appeal, aesthetic balance, and formal harmony."[63] But, he stresses, *Blue Marble* shows an entirely new cosmology: "The comfortably enclosing spheres of the pre-Copernican cosmos are absent, as is evidence of the regular technical motion of a Newtonian cosmos or of the relative space-time of Einsteinian physics."[64] The photographs combine the familiar imperial globe with an unthinkable new cosmology, visualizing the tensions between global and planetary discourse that characterize post-1968 attempts to imagine human collectives.

Astronauts and cosmonauts, who literally assume Cosgrove's Apollonian perspective, describe not so much a sense of mastery as a sublime experience that apparently negates the human through extreme scalar contrast. Frank White calls this phenomenon "the overview effect," defined as "a cognitive shift in awareness" that invokes "the experience of seeing firsthand the reality that Earth is in space, a tiny, fragile ball of life, 'hanging in the void,' shielded and nourished by a paper-thin atmosphere." This experience prompts new perspectives "on the planet and humanity's place in the universe," resulting in "a profound understanding of the interconnection of all life, and a renewed sense of responsibility for taking care of the environment."[65] For many astronauts and cosmonauts, the overview effect is a poetic experience. Astronaut Michael Collins, for one, wished for a poetic rendering of the experience: "The pity of it is that so far the view from 100,000 miles has been the exclusive property of a handful of test pilots, rather than the world leaders who need this new perspective, or the poets who might communicate it to them."[66] This scalar perspective shift is a favorite trope of poets writing of Earth in this period, as they tap into the scale intensity of lyric forms and figures such as

---

[61] Cosgrove, *Apollo's Eye*, 235.   [62] Cosgrove, *Apollo's Eye*, 262–3.
[63] Cosgrove, *Apollo's Eye*, 260.   [64] Cosgrove, *Apollo's Eye*, 278.
[65] Frank White, *The Overview Effect: Space Exploration and Human Evolution*. (Reston, VA: American Institute of Aeronautics and Astronautics, 2014), 2.
[66] Michael Collins, *Carrying the Fire* (New York: Farrar, Straus and Giroux, 1974), 471.

line, stanza, metaphor, and apostrophe that capture both the neat, measurable symmetry of the global and the unthinkability of the planetary.

Robert Poole, the definitive historian of *Earthrise*, gravitates toward poetry in his history of the photograph, frequently highlighting astronauts who meditate on poetry in orbit or wrote poems about Earth following their trips to space. Alfred Worden of Apollo 15 wrote poetry to work through his experience of seeing Earth from the moon. In "Hello Earth," printed in the NASA publication *Apollo Over the Moon* alongside photographs of Earth, Worden writes, "Quietly, like a night bird, floating, soaring, wingless / We glide from shore to shore, curving and falling but not quite touching; / Earth: a distant memory seen in an instant of repose."[67] Russell Schweickart, in discussing the emotional and spiritual experience of a spacewalk and seeing Earth revolve beneath him, associated the experience with lines from E. E. Cummings—"a blue true dream of sky; and for everything which is natural which is infinite which is yes"—which he felt the experience had made "a part of me somehow."[68] Having walked upon the moon, Buzz Aldrin read a passage from Psalm 8 on the voyage back to Earth. At one point, Poole himself is swept away and anachronistically makes an Earthrise moment of Robert Frost's desire in "Birches" (1916) to "get away from Earth awhile, / and then come back to it and begin over."[69]

NASA commissioned two poets who wrote frequently of the *Earthrise* and *Blue Marble* images: Archibald MacLeish and James Dickey. Both former fighter pilots invested in the national uses of poetry (MacLeish invented the institution of U.S. poet laureate, which Dickey held); they embraced the task of lyricizing the planet.[70] Both Dickey and MacLeish draw on their experiences as fighter pilots to channel the Apollonian gaze in verse. Dickey linked imperial aviation to the space program in public statements, and his moon landing poem appears in the same issue of *Life* that nostalgically links the historic 1927 flight of Charles Lindbergh to the crew of Apollo 11.[71] Poole attributes MacLeish's vision of "universal brotherhood" to his experience as a pilot: both astronomers and poets, Poole writes, were "influenced by the humanistic tradition of internationalist idealism which had come to maturity

---

[67] Poole, *Earthrise*, 106.
[68] Quoted in Poole, *Earthrise*, 161. The poem from which Schweickart recites is E. E. Cummings's "i thank You God for this most amazing."
[69] Poole, *Earthrise*, 163.
[70] Some of Dickey's commissioned work is notably ambivalent, as Goldstein discusses in "'The End of All Our Exploring': The Moon Landing and Modern Poetry," *Michigan Quarterly Review* 18:2 (1979): 192–216.
[71] James Dickey, "Off to the Moon," *Life*, 4 July 1969.

during the Second World War and after," with its "global vision of peace." This vision was realized, ironically, in the wartime perspective of the "airman's vision" critiqued by Auden. Poole explains,

> MacLeish's generation of idealists...had been intoxicated between the wars by the airman's vision of the world from above, free and open, at once modern and God-like. In a swords-into-plowshares movement, long-range air travel with its potential for peaceful interchange was the product of the bomber technology of the war, just as space travel was to be the product of the intercontinental ballistic missile technology of the Cold War.[72]

Commercial aviation inherited the imperial sprit of its wartime technological origins. The first commercial airline, the UK's aptly named Imperial Airways, promised the Apollonian view to citizens who could afford to fly to South Africa, India, and other places in the British Empire. In the U.S. jet age, dated to roughly the midpoint of Project Apollo in 1965, major American airlines like Pan Am and United have used the globe as their icon, rebranding imperialism as globalization.[73] Proposed commercial flights to the moon in the twenty-first century extend the reach of the global citizen into space. In the age of Apollo, NASA used Frank Sinatra's 1964 remake of Bart Howard's 1954 "Fly Me to the Moon" both to kindle the romance of space exploration and suggest that the cosmos would soon be in casual reach for all Americans.

This was the cultural moment that produced *Earthrise* as an American nationalist image. Cosgrove points to the nationalist logic that underlies these photographs' sense of uniting and saving the world, discussing how *Earthrise* and similar images suggest "visual confirmation of American democracy's redemptive world-historical mission, namely, to realize the universal brotherhood of a common humanity."[74] When *Time* featured *Earthrise* on its cover, the photo was reoriented to convey this ideology. Indeed, while the astronauts originally saw the Earth appear to the right side of the Moon and Anders, for his part, keeps his framed copy in its original orientation, the photograph is almost always displayed sideways to depict a rising Earth—or, as *Time* put it, "Dawn."[75] The reorientation of the snapshot in the press intensifies its

---

[72] Poole, *Earthrise*, 40–1.

[73] In her work on poetry of passenger flight, Marit MacArthur dates the rise of the jet age to 1965, "the year United Airlines launched its 'Fly the Friendly Skies' advertising campaign," prompting poets to begin to explore the complexities and contradictions of globalization from the perspective of air travel. See Marit MacArthur, "One World? The Poetics of Passenger Flight and the Perception of the Global," *PMLA* 127:2 (March 2012): 264–82, here 265.

[74] Cosgrove, *Apollo's Eye*, 260. [75] Poole, *Earthrise*, 28–30.

Figure 1.3 "Earthrise" (1968) in its original orientation, https://www.flickr.com/photos/nasacommons/9460163430/.

imperial message. As published, *Earthrise* offered a powerful Cold War message proclaiming the rise of an American-led utopian era of worldwide democracy (see Figure 1.3).

1.4 Lyricizing the Planet

MacLeish wrote about *Earthrise* twice in civic contexts, using the awe-invoking apostrophic "*O*" of Earth to emphasize the wonder of American democracy. When the *New York Times* published "Riders on the Earth" on the front page of their feature about *Earthrise* in 1968, the essay created planet Earth in an image of white American heteromasculinity: "To see the Earth as it truly is, small and blue and beautiful in that eternal silence where it floats, is to see ourselves as riders on the Earth together, brothers on that bright loveliness in the eternal cold—brothers who know that they are truly brothers."[76] MacLeish drew on the photograph again in 1969 when the editors of the same newspaper commissioned him to write a moon landing poem. While the poem ostensibly addresses the Moon, its opening addresses it as a "wanderer," a translation of the Greek word that gives English "planet":

> Presence among us,
> wanderer in the skies,
>
> dazzle of silver in our leaves and on our
> waters silver,
>
> O
>
> silver evasion in our farthest thought—

By the end, this pseudo-moon poem reveals itself to be an Earthrise poem, reversing apostrophic awe at the moon to awe at the Earth: "O, a meaning! // over us on these silent beaches the bright earth, // presence among us."[77] The poem ends with an unsettling collective invocation, recalling the lunar plaque that reads, "We came in peace for all mankind." The moon, in fact, gapes at Earth while orbiting it with six American flags on its surface, encircling Earth with American symbolism.

In its New Year issue of 1969—an issue that boasted "one in four Americans" would read it—*Life* magazine printed a ten-page color spread of the Apollo 8 mission and an Earthrise poem by James Dickey, "So Long" (later published as "Apollo" in his *Collected Poems*). While the title is not intended as a pun, the magazine stretches the brief poem to span three 14 × 10-inch pages of a visual story that itself documents extreme scalar distortions of space exploration. Against a space-black background, the giant white type is italicized to invoke the dreamlike temporality of extraterrestrial space. Like MacLeish's "Voyage to the Moon," "So Long" begins far away but ends with an address to Earth. It opens with "*the void*," where astronauts "*float on nothing // But procedure alone*." The poem attempts to deliver more than procedure, conveying unusual space–time in its interlinear spacing. It attempts to make the cratered moon beside it interesting, suggesting through syntactical incoherence that

---

[76] Archibald MacLeish, "Riders on the Earth," *New York Times*, 25 Dec. 1968, https://timesmachine.nytimes.com/timesmachine/1968/12/25/issue.html.
[77] John Nobel Wilford, "Astronauts Land on Plain; Collect Rocks; Plant Flag," *New York Times*, 21 July 1969, https://archive.nytimes.com/www.nytimes.com/library/national/science/nasa/072169sci-nasa.html.

the moon is too mysterious to depict in ordinary language: *"uncanny rock ash-glowing alchemicalizing the sun / With peace: with the peace of a country / Bombed-out by the universe."* The final image evokes wartime aerial bombardment, perhaps especially of Hiroshima and Nagasaki and the poem's contemporary context of Vietnam. It also suggests, as the *Earthrise* photograph does, that the Earth, if humans continue to destroy it, could become a wasteland like the moon. The poem goes on to explore the figure of lyric voicing, evoked in the flight radio that connects the astronauts and the NASA team until the astronauts pass behind the moon and lose the signal, *"the one voice / Of earth."*

The poem concludes by telling the Earthrise story, as the astronauts emerge from the dark side of the moon to find Earth. Blending prayer and poetry in apostrophic ritual, Dickey concludes by attempting to make "something" out of "nothing" through poetic creation, invoking poetry's dimensions of prayer (*"To say O God"*) and song (*"Singing with Procedure"*) to present *"The blue planet steeped in its dream / Of reality, its calculated vision shaking with / The only love."*[78]

To be fair to Dickey, it is hard to write a commissioned poem, perhaps especially about an already iconic photograph saturated with contradictory meanings. "So Long" blends elements of devotional lyric, elegy, and love poetry to deliver both political critique (that "bombed-out" country) and admiration, conveying at once violence, peace, danger, and boredom. Like the space program it was commissioned to sanctify, it has trouble finding its tone and meaning. The closest it comes is through apostrophe, evoked not only in "*O God*" and "*the one voice / Of Earth*" but also in the dreamy blue planet it displays at the end, "*shaking with / The only love."* Dickey's work is, in the end, a love poem to planet Earth. Culler routes apostrophe back to love in the moments in which a poem has no clear addressee. Sometimes, Culler writes, the "you" of the lyric is indirect and unexpressed; in these cases, it "lingers as a spectral presence, a yearning, something like love."[79] Dickey's love poem to Earth addresses "God" through the devotional "O," but really seems to want to address the unreadable, unresponsive "O" of Earth, ending on a plaintive note of unattainable love between a planet and the creatures who have left it.

---

[78] James Dickey, "So Long," *Life*, 10 Jan.1969. For an overview of Dickey's "assignment...to turn his considerable poetic talents to our exploration of space," see Irina Teveleva, "James Dickey and the Apollo Program," Washington University in St. Louis University Libraries, 18 July 2019, https://library.wustl.edu/james-dickey-and-the-apollo-program/.

[79] Culler, *Theory of the Lyric*, 243.

In addition to the commissioned works by MacLeish and Dickey in the *New York Times* and *Life*, the *National Geographic* and the *Times* included lines of verse taken out of context as captions for planet Earth photographs. The practice was widespread enough to have attracted notice by historians and literary critics. Writing of periodicals' uses of poetry in the *Earthrise* moment, Joe Luna proposes that mainstream magazines used "the invocation of poetic truth to sanction and confirm a public spectacle."[80] Luna leaves the matter there, and he certainly has a point. The mass media sought a particular kind of poem, an old-fashioned, highly lyrical composition that would treat the missions not with boredom or skepticism but, rather, with respect and gravitas. They wanted, in other words, the lyric in its ceremonial and ritualistic function, the idea of lyric as much as a specific poem.[81] One function of lyricizing *Earthrise*, then, was to celebrate and immortalize an imperial feat. The magazines employ what Ramazani calls poetry's "long-memoried" forms against the fleeting time of the breaking news story and the disposable newspaper.[82] By including poems with their *Earthrise* coverage, in other words, magazines transformed fleeting news into the timelessness of lyric reading.[83] Lyricizing a public act in a mainstream magazine takes the planet out of discursive circulation and suspends it for contemplation, elongating a fleeting national event. Yet the eerie temporality and modes of address in these poems channel not only lyric's uses to immortalize an empire but also its subversive, estranging properties.

The *National Geographic* was the first periodical to print a color photograph of Earth from space, a little-known, grainy photograph that preceded *Earthrise* by two years. It appeared in a 1966 issue with a caption from George Meredith's Italian sonnet "Lucifer in Starlight" (1883): "Above the rolling ball in cloud part screen'd."[84] Meredith's octave assumes Cosgrove's dreamy Apollonian gaze, depicting Satan rising above Earth and contemplating its topography from the African continent to the Arctic. The sestet depicts Lucifer's literal and spiritual fall. Lucifer's journey is that of the Earthrise era,

---

[80] Joe Luna, "Space / Poetry," *Critical Inquiry* 43:1 (Autumn 2016): 120.

[81] In *Theory of the Lyric*, Culler describes the ritualistic function of lyric figures: "In foregrounding the lyric as act of address, lifting it out of ordinary communicational contexts, apostrophes give us a ritualistic, hortatory act, a special sort of event in the lyric present" (213).

[82] Jahan Ramazani, *Poetry and Its Others: News, Prayer, Song, and the Dialogue of Genres* (Chicago, IL: University of Chicago Press, 2014), 103, 67.

[83] Michael Warner argues that "Lyric speech has no time: we read the scene of speech as identical with the moment of reading. Public speech, by contrast, requires the temporality of its own circulation." Warner, *Publics and Counterpublics* (New York: Zone Books, 2010), 81.

[84] Quoted in Kenneth F. Weaver, "Historic Color Portrait of Earth from Space," *National Geographic* 132:5 (1967): 726.

away from the planet and then back to it. The planet as a "rolling ball" evokes the earliest documented contemplation of the spherical Earth, from Plato: "The true Earth, if one views it from above, it is said to look like those twelve piece leather balls, variegated, a patchwork of colours of which our colours are, as it were, samples that painters use."[85] The balance of these images, in both form and color, finds its analogue in patterned verse. Plato's aesthetically harmonious geocentric cosmos gave way to Judeo-Christian cosmology that locates God in heavenly spheres above a central Earth. "Above the rolling ball in cloud part screen'd" therefore appeals to classical aesthetics and the Christian cosmogenic narrative.[86] Meredith's rhyme of "Awe" and "law" restores cosmic harmony through marveling at God's perfect creation, sanctified by the laws of the sonnet form in tying together the sestet:

> Soaring through wider zones that pricked his scars
> With memory of the old revolt from Awe,
> He reached a middle height, and at the stars,
> Which are the brain of heaven, he looked, and sank.
> Around the ancient track marched, rank on rank,
> The army of unalterable law.[87]

Meredith's harmonious sonnet seems to arrest time in its insistent repetition of *now*, invoking the immutability of the ancient Greek cosmic spheres. The poem's overall metrical and sonic balance mimics the orderly, eternal cosmos Meredith inherits from Milton's *Paradise Lost* ("Their starry dance in numbers that compute")—that steady iambic pentameter suggesting divine order. Only Lucifer's moves are irregular: the "dark planet" he becomes does not follow any measurable orbit, certainly not the "ancient track" of the zodiac the steady stars trace, but rather careens over Africa and the Arctic and then through the "wider zones" that refer to a steady classical cosmology of celestial spheres. In the face of the steady celestial plane—that "army of unalterable law"—Lucifer's orbit breaks, and he sinks back to Earth.

The poem draws on the Apollonian gaze even as it foregrounds the unusual ways in which poems use language. Even as the poem thematizes the ethics and aesthetics of balance and harmony, it foregrounds the strangeness of its

---

[85] Quoted in Cosgrove, *Apollo's Eye*, 54.
[86] For a discussion of this Christmas Eve broadcast in the context of *Earthrise*, see Poole, *Earthrise*, 8.
[87] George Meredith, "Lucifer in Starlight," in Laura Emma Lockwood, *Sonnets, Selected from English and American Authors* (Boston, MA: Houghton Mifflin Co., 1916), 175.

own language in unusual metaphors (that "brain of heaven" tangling scientific and religious discourse) and inverted syntax. The *National Geographic* caption, "Above the rolling ball in cloud part screen'd," exalts the image of Earth, but it also estranges it, rendering the planet hard to see in language; that "rolling ball," moreover, is occluded by clouds. The line conveys the strangeness of confronting Earth as an otherworld through this opacity and through the alien perspective of Satan, an otherworldly visitor.

No periodical better demonstrates the estranging quality of poetry in the Earthrise moment than the *Times*, which published *Earthrise* on its cover with lines from John Keats's "On First Looking into Chapman's Homer": "then I felt like some watcher of the skies when a new planet swims into his ken."[88] (The *Times*, as Luna notes, misses the line break between "skies" and "when."[89]) The poem was written in response to F. W. Herschel's discovery of Uranus in 1781.[90] Thus, including it here frames Earth as a newly discovered planet, offering an impossible otherworldly encounter with the planet to which our species is attached. The *Times*, in fact, brings the image into the proximity of fantasy and science fiction. Robert J. Tally observes that Earth photography evokes a sense of estrangement that reorganizes an abstracted cosmopolitan perspective; with *Earthrise*, people saw the planet "as a strange *otherworld*, as if for the first time. Arguably, in the contemplation of this image and of its ramifications, a planetary consciousness emerged."[91] Tally usefully brings the planetary into the realm of science fiction and fantasy, both of which he understands as literatures of "cognitive estrangement." The lyric also does the work of cognitive estrangement, as science fiction theorist Darko Suvin argues; he evokes Shelley's idea of lyric poetry—that which "makes familiar objects be as if they were not familiar"—when he writes that poetry "bring[s] out the unexpected from the encrusted."[92] To keep the moment of seeing the planet from being reduced to the stuff of "mere" science fiction, Apollo-era institutions brought in the lyric to do the work of science

---

[88] Pearce Wright et al., "The Colour of Space: Earth Photographed by Man from over the Moon," The *Times*, 6 Jan. 1969, 1.

[89] Luna, "Space / Poetry," 120.

[90] In 1811, Keats received a copy of *An Introduction to Astronomy* by John Bonnycastle, a widely read text of popular astronomy. For the influence of this book on Keats, see Dometa Wiegand Brothers, *The Romantic Imagination and Astronomy: On All Sides Infinity* (New York: Palgrave Macmillan, 2015), 99; Anna Henchman, *The Starry Sky Within: Astronomy and the Reach of the Mind in Victorian Literature* (Oxford: Oxford University Press, 2014), 28.

[91] Robert T. Tally Jr., "Beyond the Flaming Walls of the World: Fantasy, Alterity, and the Postnational Constellation," in Elias and Moraru, *The Planetary Turn*, 196.

[92] Percy Bysshe Shelley, *A Defense of Poetry* (Boston, MA: Ginn & Co., 1840). Darko Suvin, *Defined by a Hollow: Essays on Utopia, Science Fiction and Political Epistemology* (Bern, Switzerland: Peter Lang, 2010), 22.

fiction, drawing on its elite generic standing to legitimize and venerate an otherworldly encounter with our own world. Arrested in lyric time and presentation on the disposable pages of mass-produced newspapers, planet Earth stares at its readers in the hard-to-think time of suspended apostrophic address. In its reorganizing of time and space, its use of language against the grain of typical expression, and its odd ways of confusing the coordinates of "I" and "thou," lyric poetry conveys the oddness of reading an image that ought to be impossible for planet-bound creatures to encounter.

1.5 Dreaming of Other Worlds

On Christmas Eve 1968, five lunar orbits after Anders snapped *Earthrise*, the crew of Apollo 8 made a telecast. Broadcast live on five continents, it began with Borman's explanation of the camera's perspective: "We showed you first a view of the Earth as we've been watching it for the past 16 hours. Now we're switching so that we can show you the moon that we've been flying over at 60 miles altitude for the last 16 hours." Borman's language is telling; though the crew has been in lunar orbit, "flying over" the stated objective of Project Apollo—the moon—they have "been watching" the now-distant Earth the whole time. As the broadcast continues, Borman gives insight as to why the astronauts' gazes did not favor the moon. "The moon is a different thing to each of us," Borman says: "I know my own impression is that it's a vast, lonely, forbidding type existence, great expanse of nothing...and it certainly would not appear to be a very inviting place to live or work." Lovell echoes Borman: "The vast loneliness up here of the moon is awe inspiring," he explains, "and it makes you realize just what you have back there on Earth." Only Anders seems interested in the lunar surface. He conveys a fascination with "lunar sunrises and sunsets" that "bring out the stark nature of the terrain." Anders does his best to describe the lunar landscape as it slips beneath the camera's view and his colleagues interrupt with advice. Interestingly, both Borman and Lovell mention color. Borman reminds Anders that their footage is "not in color," while Lovell interrupts Anders's description of the lunar horizon—"a vivid, dark line"—with a revision: "Actually, I think the best way to describe this area is a vastness of black and white, absolutely no color."[93]

---

[93] Toward the end of the broadcast, Anders takes the party line: "I hope all of you back down on the earth can see what we mean when we say that it is a rather foreboding horizon, a dry rather dark and unappetizing place."

64   LYRIC POETRY AND SPACE EXPLORATION

As Apollo 8 "approach[ed] the lunar sunrise" and the lunar surface brightened on televisions across the planet, Anders switched from description to announcement:

For all the people back on earth, the crew of Apollo 8 has a message that we would like to send to you.

In the beginning, God created the Heaven and the Earth. And the Earth was without form and void, and darkness was upon the face of the deep. And the spirit of God moved upon the face of the waters, and God said, "let there be light." And there was light. And God saw the light and that it was good, and God divided the light from the darkness.

Lovell continued, reading from Genesis 1:4–8 about the firmament that, like the "vivid, dark line" Anders described on the lunar horizon, divided what was above the line from what was below, the heavens from the waters. Borman picked up the demarcation of land and sea in verses 9–10: "And God called the dry land Earth. And the gathering together of the waters called the seas. And God saw that it was good." As commander of the mission, it was Borman who had first been charged with saying "something appropriate" on the Christmas Eve telecast. He first tried writing something of his own, but he abandoned that approach and enlisted the help of others, including Joseph Laitin, a governmental official outside of NASA. When Laitin also had trouble, his wife, Christine, suggested the passage from Genesis.[94]

While Genesis 1 is not poetry in the strictest sense—it has the characteristics of Hebrew history rather than Hebrew verse—the text is widely seen as poetic, and its use on Apollo 8 coincides with NASA's deployment of lyric to convey the experiences of space. The timing of the broadcast—Christmas Eve—aligns American space exploration with both the "chosen people" of Genesis and the messianic tradition Christians read as the teleological purpose of those "chosen" and the Creation itself, with Jacob and the "Star out of Jacob."[95] Lyric often reinforces the Christian redemptive overtones of

---

[94] J. Y. Smith, "Christine Laitin Dies at 65," *Washington Post*, 6 Apr. 1995, https://www.washingtonpost.com/archive/local/1995/04/06/christine-laitin-dies-at-65/62c0b636-aee2-479d-81a5-5ea52179e5cc/.

[95] See Genesis 28:10–17 and Numbers 24:17. John Aaron, who served at Mission Control for Apollo 8, recollects the moment: "I don't think anybody knew they were gonna do that. The hair stood up on the back of my back. The first impression I had was, how appropriate." Jerry Bostick, who also served at Mission Control, calls the experience "one of the most memorable things in my life.... What could be better than having the first human beings, Americans, circling the moon on Christmas Eve, and they read the story of Creation from Genesis? I mean, it brought tears to my eyes." "Apollo's Daring Mission," *NOVA*, PBS, 1968.

messianic rhetoric, tying that redemption to the American narrative. This equation through lyric has a powerful impact: going into space means becoming God. Indeed, this deployment of Genesis overshadows the Greek god for whom Project Apollo was named: Apollo may have pulled the sun with his Chariot, but God—now directly connected to American ingenuity—created light itself. Borman transitioned without pause from the text of Genesis to his closing: "And God saw that it was good. And from the crew of Apollo 8, we close with good night, good luck, a Merry Christmas and God bless all of you—all of you on the good Earth." His repetition of the words *and* and *good* enmeshes the message with that of Genesis, both in content and in perspective.

Though the crew of Apollo 8 read, ostensibly, as private citizens, both NASA and the U.S. benefitted from the lyricization of earth and its scalar shift in perspective.[96] This shifting lyricization continued with the proliferation of *Earthrise*. In 1969, the U.S. Post Office combined these famous moments from the Apollo 8 mission, releasing a stamp of *Earthrise* with the words, "In the beginning, God...." This pre-Copernican Christian cosmogenic narrative has long been attached to Euro-American ideas of the U.S. as the "New World," as in John Locke's declaration, "in the beginning all the world was America."[97] The Christian redemptive dimensions of the American Earth are frequently outsourced to lyric, which secularizes the ritualistic functions of prayer. Ramazani suggests that prayer and poetry live close together in "their rhetorical stance as apostrophic discourses," noting "the similarities between poetic apostrophe and the rhetorical structure of prayer" in an often inward address to imagined listeners.[98] Anders's response to the view that would be translated into the photograph—"Oh my God!"—echoes in the apostrophe of Dickey's first commissioned poem for *Earthrise*: "Oh God, behold / the blue planet...."[99] This lyrical-devotional-nationalist tangle brought Christian creation and redemption to bear on the U.S.'s resurrection of the ailing globe—and planet. The Apollonian perspective draws on lyric to promote the idea of an expansive, transcendent, even godlike American spirit on the global stage, naturalizing and even immortalizing U.S. power through metaphors of deep time. Visualizing Earth from extraterrestrial space also imbues American

---

[96] David Welna, "Space Force Bible Blessing at National Cathedral Sparks Outrage," *NPR*, 13 Jan. 2020, https://www.npr.org/2020/01/13/796028336/space-force-bible-blessing-at-national-cathedral-sparks-outrage.
[97] John Locke, *Locke: Two Treatises of Government*, ed. Peter Laslett (Cambridge: Cambridge University Press, 1960), 99.
[98] Ramazani, *Poetry and Its Others*, 128–39.   [99] James Dickey, "So Long," 8.

power with divine right by invoking an aesthetic-devotional tradition reaching from the sun-poet god Apollo to the astronauts who read the Christian cosmogenic narrative in lunar orbit on Christmas Eve, 1968. At the same time, the perspective revealed destabilizes national borders and even ideas of the harmonious globe that the U.S. sought to consume within its own ever-widening borders.

Across national, global, and planetary discourses, Earth photographs summon lyric's optative and utopian qualities. These photographs are political in their very attempts to bracket off the horrors of the world. For Theodor Adorno, lyric is at its most political when it seems the most withdrawn from society, recording the collective ills that drive it to loneliness and alienation. In their lyricization and lyrical qualities, these photographs register the dystopian forces that make the planet seem to want to turn its back on the world. "Steeped in its dream / of reality," the lyrical planet expresses above all, as Adorno wrote of poetry, "the dream of a world in which things would be different."[100] Subsequent chapters will consider the lyric's embeddedness in and resistance to imperial power, as well as its capacity to create other worlds when there is no way out of this one.

---

[100] Theodor Adorno, "On Lyric Poetry and Society," in *Notes on Literature*, 1, (New York: Columbia University Press, 1991), 39–40.

2
"Galaxies of Women"

Containment Culture and the Queer Astronomical Lyric

Elizabeth Bishop reminded James Merrill of a planet. In an elegiac tribute, Merrill recalls, "The whitening hair grew thick above a face each year somehow rounder and softer, like a bemused, blue-lidded planet, a touch too large...for a body that seemed never quite to have reached maturity."[1] In Merrill's queer portrait, Bishop's body defies the "natural" laws of aging and inverts the national law of what Jack Halberstam identifies as "reproductive temporality": her hair is "whitening" as her face becomes more youthful, ending with an immature body that did not reproduce.[2] Ten years later, Merrill again described Bishop as planet Earth. He styles her posthumously published letters as messages sent from beyond the world:

> The anguish coming only now to light
> In letters like photographs from Space, revealing
> Your planet tremulously bright through veils
> As swept, in fact, by inconceivable
> Heat and turbulence—[.][3]

The cloud-veiled planet, like Bishop herself, is both brilliant and unfathomable. In pairing the word "revealing" with "veils" (the sonic partner of the longer word that anagrammatically conceals its letters), the elegy links the glowing, cloud-wrapped planet to epistemological limits that convey Bishop's complexity. As similes for Bishop's letters, these photographs suggest the limits of both textual and personal readability. Merrill's opaque, queer reading of Bishop offers this poet—who was notoriously difficult to classify in life or in art—the right to inscrutability.

---

[1] James Merrill, "Elizabeth Bishop, 1911–1979," in *Elizabeth Bishop and Her Art*, ed. Lloyd Schwartz and Sybil P. Estess (Ann Arbor: University of Michigan Press, 1983), 259.

[2] Jack Halberstam, *In Queer Time and Place: Transgender Bodies, Subcultural Lives* (New York: New York University Press, 2005), 4.

[3] James Merrill, *Collected Poems* (New York: Knopf, 2002), 666–7.

This chapter extends Chapter 1's insights about how the planet became a lyrical object into a consideration of the larger space age's impacts on mid-century lyric poetry. I turn to the careers of two seemingly unalike poets—Bishop and Adrienne Rich—to reveal how space age politics and aesthetics inflected the development of a mid-century queer lyric "I." Both use astronomy to subvert Cold War surveillance as they wrote in a period that equated homosexuality with treason. Attending to how the foreign policy of containment impacted American domestic life, I will consider how containment culture operated through a central paradox of "containing" and "expanding," with far-flung extraterrestrial exploration at once the antithesis of and the fullest realization of containment. Containment also spoke to formal debates in mid-century poetry about "breaking out" of restrictive poetic forms to embrace the authentic self in the confessional mode, as Christopher Grobe has demonstrated.[4] Bishop and Rich rose to prominence at the height of the confessional movement but were not confessional poets. Instead, their work uses astronomy to explore lyric subjectivity while subverting the apparent transparency of confession. I consider Rich's turn from formalism to free verse poetry through her series of astronomical poems, beginning in 1953 with "The Explorers" (which contemplates, in heroic couplets, the Earth seen from the moon) and ending with her 1968 experimental poem "Planetarium," celebrated as her first poem to achieve a politically inflected but deeply personal subjectivity. I then trace Bishop's astronomical poetics from her Sapphic love poems under the night sky of Brazil to her poem "In the Waiting Room," where her speaker recognizes that she is "an *I*" among others after feeling "the sensation of falling off / the round, turning world / into cold, blue-black space."[5] Bishop and Rich both have hard-to-place relationships to mid-century lyric, and to one another; neither allies nor foils, they watched one another's work with detached interest. This chapter attempts to understand why these poets, working in such dissimilar aesthetic and political modes that that they often missed one another's points, turned to astronomy in similar ways during the 1950s–1960s as they developed opaque lyric voices. Bishop's and Rich's astronomical lyrics, which abound with optical metaphors and techniques, channel Cold War surveillance into strategies of opacity as they develop queer poetics in a cultural moment that actively dehumanized queer people in the service of American foreign interests.

---

[4] Christopher Grobe, *The Art of Confession: The Performance of Self from Robert Lowell to Reality TV* (New York: New York University Press, 2017), 28–9.

[5] Elizabeth Bishop, *Geography III* (New York: Farrar, Straus and Giroux, 1976), 6. Further references are noted in the text as G.

2.1 The Space Age and the "Breakthrough" Narrative

In the years their country began blasting rockets into space, American poets started embracing the expressive "I" like never before. Squarely between the launch of Sputnik 1 in 1957 and John F. Kennedy's 1962 vow to put a man on the moon, Robert Lowell's *Life Studies* (1959) galvanized the confessional movement. In his acceptance speech for the National Book Award, Lowell influentially distinguished between the "cooked" verse of well-mannered formalists and the "raw" sensibilities of the Beats. In his move away from his earlier tightly metrical and rhymed verse in *Life Studies*, he found the method for a therapeutic "breakthrough back into life" out of the cage of poetic form.[6]

The Cold War's dominant metaphor of containment drove astronauts to break out of the closed terrestrial globe and poets to break free of the stifling New Critical poem in the same years. The geopolitical policy named by George F. Kennan to describe the U.S.'s strategy to contain the spread of communism, containment, as Alan Nadel has demonstrated, was not just a foreign policy: it also described U.S. domestic values and codes of behavior. Nadel argues that "cold war America asserted the claim to global authority in a narrative that permeated most aspects of American culture"[7]—from space exploration to domestic life to the arts. Elaine Tyler May locates containment not only within U.S. domestic borders but specifically within the home itself, reimagined as a site of American freedoms through marriage and reproduction that provided a fortress against international conflict.[8] Indeed, many women confessional poets figure their "breakthroughs" by exposing the constraints, rather than the freedoms, of the Cold War nuclear family. "Breakthrough," Grobe suggests, "is containment's natural complement—a world-shaping metaphor of equal and opposite force."[9] Tracing the origin of the term "breakthrough" to the World War I battlefield, Grobe argues that the breakthrough seeks a "violent rupture" with the repressive containment culture that pervaded both U.S. foreign policy and domestic life.[10] As Grobe puts it, "no matter how ideologically bland they may seem to us now,

---

[6] Robert Lowell, "Robert Lowell, The Art of Poetry No. 3," *Paris Review* 25 (Winter–Spring 1961), https://www.theparisreview.org/interviews/4664/the-art-of-poetry-no-3-robert-lowell.

[7] Alan Nadel, *Containment Culture: American Narratives, Postmodernism and the Atomic Age* (Durham, NC: Duke University Press, 1995), 4.

[8] Elaine Tyler May, *Homeward Bound: American Families in the Cold War Era* (New York: Basic Books, 1988). May writes, "in the early years of the cold war, amid a world of uncertainties...the home seemed to offer a secure private nest removed from the dangers of the outside world" (May, *Homeward Bound*, 3).

[9] Grobe, *The Art of Confession*, 29.   [10] Grobe, *The Art of Confession*, 27–9.

confessionalists saw themselves this way: as the enemy within, lurking in the suburban middle class, plotting a breakthrough past the forces of cultural containment." As these poets languished in the suburbs plotting their escape through free verse, the Kennedy administration was insisting that space exploration was vital to the success of containment. As Kennedy put it, "Control of space will be decided in the next decade. If the Soviets control space they can control earth, as in past centuries the nation that controlled the seas dominated the continents.... To insure peace and freedom, we must be first."[11] Only by breaking out of the borders of the world itself would it be possible to contain communism and enforce values of democracy and capitalism. To control the globe by leaving it: space exploration highlighted a paradox at the heart of the containment metaphor, as extraterrestrial exploration became at once the antithesis and the fullest realization of containment. The space program supported containment, while the confessionals challenged its focus on conformity. Yet many mid-century poets end up where the space program did: claiming freedom and the values of individualism through breaking out of an enclosed form, be it the well-wrought urn of the New Critical poem or the bounded sphere of the terrestrial globe. Ironically, this turn to free verse symbolically endorsed containment policy by representing the very freedoms and self-expressiveness U.S. institutions were promoting within and beyond the nation's borders.

The space age "I," from the astronaut to the confessional poet, had a peculiar Cold War inflection, caught up in national paranoia about communist totalitarianism. In the 1950s and especially the 1960s, free verse became aligned with personal freedoms and formal verse with conformity. Free verse, of course, predates the Cold War. Whitman "freed up" the iambic line as early as the 1840s in the name of democracy, and Ezra Pound and T. S. Eliot extended *vers libre* in the 1910s through their readings of French poets.[12] Moreover, it is a commonplace that "free verse" is not actually "free"; as Donald Hall memorably put it, "the *form* of free verse is as binding and as liberating as the *form* of a rondeau."[13] But free verse took on Cold War political symbolism. The "well-made poem" of Lowell's "cooked verse" became synonymous with conservative elements of containment culture, despite the

---

[11] John F. Kennedy, Moon Speech, Rice Stadium, Rice University (12 Sept. 1962), https://er.jsc.nasa.gov/seh/ricetalk.htm.

[12] Ezra Pound, *ABC of Reading* (Toronto, ON: Penguin Books, 1934); T. S. Eliot, "Reflections on Vers Libre," in *The Complete Prose of T.S. Eliot: The Critical Edition*, ed. Ronald Schuchard (Baltimore, MD: Johns Hopkins University Press, 2021), 511–18.

[13] Donald Hall, "Goatfoot, Milktongue, Twinbird," in *Goatfoot, Milktongue, Twinbird: Interviews, Essays, and Notes on Poetry, 1970–76* (Ann Arbor: University of Michigan Press, 1978), 121.

leftist politics of some formalists. The reaction against "cooked" verse registered a larger rejection of containment's emphasis on conformity.[14] Free verse came to signal American freedoms and the celebration of individualism, in much the same way as the space race was sold as freedom of choice and celebration of the heroic everyman. "We choose to go to the moon," Kennedy said in 1962, suggesting that the capitalist value of choice would lead the nation to political victory.[15] Lyric poetry got strangely mixed up in this mission. Just as Americans could "choose" to go to the moon, they could "choose" their own verse forms that best reflected their personal subjectivities and poetic ambitions. The CIA, as part of this logic, even promoted free verse in the decolonizing nations the U.S. and USSR vied for control over, funding literary festivals and journals to promote American freedoms that aligned with Kennedy's optimistic space age rhetoric.[16]

The confessional poet, like the astronaut, was an icon of individualism positioned to save the world from communism. Confessionalism took the image of the individual lyric poet and raised the stakes, demanding a break with conformity into taboo subjects and frank expressions of unique selfhood incompatible with communist homogeneity. In *Workshops of Empire*, Eric Bennett demonstrates how Cold War values of literary production supported "salvific individualism—the aesthetics of the unique person" in the fight to contain communism.[17] The rise of the New Criticism and liberal humanism in the decades before the Cold War, Bennett suggests, paved the way for the unique and bounded subject to express democratic freedoms and fulfill "the destiny of humankind" in the 1950s and 1960s.[18] Lyric poetry, with its cultural purchase on subjectivity, was especially well positioned to advertise and value the "free individual" in this climate.

So was the astronaut. Astronauts became iterations of the everyman through their individual biographies that emphasized Cold War family values of marriage and heterosexual reproduction in white suburban spaces. The

---

[14] Nadel writes, "'conformity' became a positive value in and of itself. The virtue of conformity—to some idea of religion, to 'middle-class' values, to distinct gender roles and rigid courtship rituals—became a form of public knowledge through the pervasive performances of and allusions to containment narratives." See Nadel, *Containment Culture*, 4.

[15] John F. Kennedy, "Address at Rice University on the Nation's Space Effort" (12 Sept., 1962), https://er.jsc.nasa.gov/seh/ricetalk.htm.

[16] See David Caute, *The Dancer Defects* (Oxford: Oxford University Press, 2003), 3; Andrew N. Rubin, *Archives of Authority: Empire, Culture, and the Cold War* (Princeton, NJ: Princeton University Press, 2012), especially 47–73; and Nathan Suhr-Sytsma, *Print, Poetry, and the Making of Postcolonial Literature* (Cambridge: Cambridge University Press, 2017), especially 60–74.

[17] Eric Bennett, *Workshops of Empire: Stegner, Engle, and American Creative Writing During the Cold War* (Iowa City: University of Iowa Press, 2015), 32.

[18] Bennett, *Workshops of Empire*, 36.

barriers would-be astronauts of color had to overcome to access these spaces contributed to the overwhelming whiteness of the American space program—and of the confessional movement, full of poets trapped in the suburbs where they could expose the psychological damages of white middle-class life.[19] Confessional poets wrote from and of the turmoil that beset many astronauts' submerged post-space biographies, with many going on to battle substance abuse and mental illness.[20] But publicly these two icons of individualism lived different lives: confessionals expressed "universal" (white middle-class) suffering through their particularized pain, while astronauts celebrated the success of American individualism. Both public personas, however, embraced an expansive space age "I" announcing itself across the universe.

Yet the critical story of confessional verse, in its focus on the amplified self, often misses the opacity at the heart of confessionalism. Deborah Nelson's work on constitutional privacy and confession paves the way for such considerations. She finds in confessional verse a white masculine desire for privacy, as it was being increasingly violated by the citizen surveillance of the Cold War (which was, ostensibly, a way to secure and protect that very privacy). Confessionalism is not about self-revelation for its own sake, but rather about regulating one's privacy, which, Nelson contends, "is available to the extent that its violations reinvent it."[21] Modulating privacy through confession was especially important for women poets, Nelson argues, as they were subject to constant exposure within and beyond the home. Women confessional poets generate the privacy denied to them through confession. This act depends on an "I" that gives the impression of concealing hidden depths.

In her extraterrestrial poems, for example, Sylvia Plath frequently gives the illusion of offering up the self, only to make it inaccessible. In "Wuthering Heights," the speaker watches sheep who watch her back, rendering her unreadable: "The black slots of their pupils take me in. / It is like being mailed into space, / A thin, silly message."[22] "Take me in" suggests that the sheep are bringing the speaker into their interiority—whatever that is, for sheep. Entering their interiority is as strange and futile as being mailed into space.

---

[19] For an overview of how the control of suburban white spaces mapped onto white fantasies of controlling extraterrestrial space, see Lynn Spigel, *Welcome to the Dream House: Popular Media and the Postwar Suburbs* (Durham, NC: Duke University Press, 2001), especially 145.

[20] See Buzz Aldrin, *Magnificent Desolation: The Long Journey Home from the Moon* (New York: Three Rivers Press, 2010); Scott Kelly, *Endurance: My Year in Space, A Lifetime of Discovery* (New York: Knopf, 2017).

[21] Deborah Nelson, *Pursuing Privacy in Cold War America* (New York: Columbia University Press, 2002), xx.

[22] Sylvia Plath, "Wuthering Heights," in *Sylvia Plath: The Collected Poems* (New York: HarperCollins, 2008), 167. Further references are noted in the text.

The act of mailing, like lyric address, presupposes a relationship between an "I" and a "you," a sender and a recipient of a message. In this passage, the communication circuit breaks down. The sheep's "hard, marbly baas" are as indecipherable to her as the language of poetry is to them; on at least one end of the communicative circuit is an interpretive void.

In "A Birthday Present," Plath again uses extraterrestrial figures within a confined domestic space to create effects of opacity. The present of the title becomes a metaphor for the inaccessible self wrapped in the domestic packaging of marriage and childrearing: "What is this, behind this veil, is it ugly, is it beautiful?" the speaker wonders while "[m]easuring the flour, cutting off the surplus, / Adhering to rules, to rules, to rules." The couplets of the poem evoke the domestic rules of measurement and repetitive duties, even as the variable line lengths keep the poem from adhering entirely to the neatness of "edges" and "rules." The poem ends by ripping off the wrapping around the self: "Only let down the veil, the veil, the veil," the speaker implores: "And the knife not carve, but enter // Pure and clean as the cry of a baby, / And the universe slide from my side" (207–8). The poem imagines the only release from domestic toil is in dying, likened to the vast and disorienting expanse of the universe. The phrasing "slide from my side" also suggests that the speaker is a kind of unbounded universe herself; if she were to cut into herself, the whole universe would spill out. The vastness of the universe waits within the wrapped gift, the enclosed kitchen, and the container of the woman's body. This is not the bounded "I" of the well-made poem, or the apparently accessible "I" imagined in the act of confession, but rather an expansive and nebulous one. In both "Wuthering Heights" and "A Birthday Present," Plath creates speakers half-veiled in cosmic unfathomability. Her confessional verse seems to offer the self transparently to a reader, only to exhibit how little that self can be tracked, classified, and known.

2.2 Cold War Surveillance and Queer Opacity

For queer poets writing within a surveillance state that attempts to monitor and control queer communities, opacity has added stakes. As satellites began to orbit the planet watchfully, the U.S. government was watching its citizens—especially for evidence of homosexuality. As Elaine Tyler May has demonstrated, containment upheld the home as a specifically heterosexual space, offering the realization of one's fullest American life through marriage and, above all, reproduction. To deviate from Lee Edelman's "reproductive futurism" was to undermine the American nationalist project.[23] "Many high-level government officials...believed wholeheartedly that there was a direct connection between communism and sexual depravity," May writes. In 1950, the year the childless, unwed Bishop completed her government role as poetry consultant in plain view of Washington,

> the Senate issued a report on the *Employment of Homosexuals and Other Sex Perverts in Government*, which asserted that "those who engage in overt acts of perversion lack the emotional stability of normal persons"....Like communists, who would infiltrate and destroy the society, sexual "perverts" could spread their poison simply by association.[24]

The Cold War surveillance state trained its gaze on homosexuals. As Foucault argues, the surveillance state relies on confession as a way of producing knowledge—especially of sexual behavior—as a means of social control. In the surveillance state, he theorizes, surveillance becomes internalized so that one is always conducting surveillance of oneself, confessing in order to be punished and forgiven.[25] Confession comes to feel liberating when in fact it has been mandated all along as a way to track, classify, and control citizens' behaviors.[26] Sex, Foucault argues, is a "privileged theme of confession," attached to the disclosure of the deepest secrets of one's authentic self.[27] Extending Foucault to challenge the exposing metaphor of coming out of the closet, de Villiers argues that "homophobia often insists on knowing rather than refusing to know about the sexuality of gay people."[28] Opacity creates epistemological impasses so that the subject cannot be mastered. De Villiers thereby defines "*queer opacity*" as "the possibility of non-meaning and non-knowledge" that works "[a]gainst the hermeneutics of sex as a field of meaning to be deciphered and interpreted."[29] Instead, de Villiers suggests, many queer thinkers and artists turn to opacity "as an alternative queer strategy or tactic that is not linked to an interpretation of hidden depths, concealed

---

[23] Lee Edelman, *No Future: Queer Theory and the Death Drive* (Durham, NC: Duke University Press, 2004), 19.

[24] May, *Homeward Bound*, 90–1.

[25] Michel Foucault, *Discipline and Punish: The Birth of the Prison*, trans. Alan Sheridan (New York: Vintage Books, 1995), 175.

[26] Michel Foucault, *The History of Sexuality, Volume 1: An Introduction* (New York: Vintage Books, 1978), 60. He observes how "the obligation to confess is now relayed through so many different points...that we no longer perceive it as the effect of a power that constrains us."

[27] Foucault, *The History of Sexuality*, 61.

[28] Nicholas de Villiers, *Opacity and the Closet: Queer Tactics in Foucault, Barthes, and Warhol* (Minneapolis: University of Minnesota Press, 2012), 2–3.

[29] Villiers, *Opacity and the Closet*, 15–16.

meanings, or a neat opposition between silence and speech."[30] Bishop and Rich develop queer strategies of opacity to thwart the logic of epistemological mastery and confession.

Merrill's elegy for Bishop, with which this chapter begins, suggests the queer opacity of her work through poetic strategies that at once conceal and reveal her. His sonnet sequence "Overdue Pilgrimage to Nova Scotia" pays homage to Bishop by returning to her birthplace and imagining the future of his own work. The poem's conceit is his pilgrimage to an ancestral point in his chosen queer poetic lineage. But when he arrives, he finds scenes that are decidedly *unheimlich* and artificial, leading the sonnet to compare many types of artifice: the "art" of "living" and of poetry, Bishop's letters to "photographs from Space," an "ESSO station" to a "shrine" of her poetry, and Bishop herself made into a work of art by the sonnet. Whenever the poem moves to "reveal" Bishop, it pulls back. The promise of "revealing / Your planet" is interrupted by a line break and the clouds covering Earth. The word "revealing" chimes with subsequent end-words evoking obscurity: "veils," "inconceivable," "it," "read," "sign." Merrill lends Bishop a level of unreadability through high artifice.[31]

Writing in a queer formalist tradition traceable to Auden, Merrill puts artifice on display to play with the association of queerness and unnaturalness, in the sense of homosexuality as an abomination against nature. In its high artifice, formalist verse has subversive queer potential. As contemporary poet Richie Hofmann, who consciously positions himself in this tradition, writes,

> *There's something queer to me about poetic form, in general.... Even though certain rules and structures of poetry can become codified as artistic law over time, poetry—or poetic utterance—remains disruptive. It remains uncategorizable.... Rhyming is about bringing things into relation, often in surprising and non-normative ways. That seems queer to me.*[32]

Throughout her work, Bishop galvanizes "rules and structures of poetry... codified as artistic law"—including meter, rhyme, and apostrophe—for their inherent "disruptive" potential. In an era that aligned formalism with conformist heteronormative suburban life, Merrill's and Bishop's formalism has

---

[30] Villiers, *Opacity and the Closet*, 3, 6.   [31] Merrill, *Collected Poems*, 248.

[32] Richie Hofmann, "Innovation in Conversation (Part IV): Speaking with Randall Mann, Richie Hofmann, Phillip B. Williams, and Chen Chen," *Kenyon Review* (Feb. 2017), http://www.kenyonreview.org/2017/02/innovation-conversation-part-iv-speaking-randall-mann-richie-hofmann-phillip-b-williams-chen-chen/.

an unexpectedly rebellious dimension. Queer opacity, then, offers a new way to understand Bishop's work, not as a conservative poetics out of touch with the experimental forms that better spoke to the crises of the period but, rather, as a subversion of transparency and surveillance. Queer opacity also offers insight into Rich's career. Armed with the surprising opacity of the confessional mode, and adding to it strategies of queer opacity, Rich and Bishop work within and against the containment logic of confessional verse to open their poetry to explorations of internal and external otherness.

Neither Bishop nor Rich quite had a coming-out-of-the closet "breakthrough" narrative in life or in art. After a visit from Rich, where Rich apparently pushed Bishop to be more forthcoming about her sexuality, Bishop delivered her famous remark, "I want closets, closets, and more closets."[33] Rich, in contrast, is often read through the breakthrough narrative, as she moved from formal to free verse while becoming an activist for lesbian rights. Rich's story, though, is not so much one of "coming out" as one of identifying a mode of relation among women that has always been in plain sight. Rich introduced the "lesbian continuum" as an antidote to the "compulsory heterosexuality" that she identified as an engine of American nationalism in its sustenance of interlocked family, religious, and state values. Rich's lesbian continuum proposes "a range—through each woman's life and throughout history—of women-identified experience," referring not just specifically to sexual desire but also to "intensity between and among women, including the sharing of a rich inner life, the bonding against male tyranny, the giving and receiving of practical and political support."[34] The lesbian continuum aims for intersectionality—Rich attempts to decenter white and middle-class experience by demonstrating how women across times, places, and races have supported one another in emotional, sexual, and economic arrangements. But its universalizing understanding of "woman" suggests the need for finessing its conceptions of difference. Christopher Spaide argues that Rich's career was driven by this determination to finesse difference within the ostensibly individual, apolitical lyric—and that ultimately, this is what makes the "confessional" a poor category through which to understand her work. Analyzing Rich's "two-decade swerve from the universal 'we' of humankind to an 'I' of empowered individuality, then to a newfound 'we' of feminist collectivity," Spaide brings to light "Rich's career-spanning effort to open up the lyric, that

---

[33] Gary Fountain and Peter Brazeau, *Remembering Elizabeth Bishop: An Oral Biography* (Amherst: University of Massachusetts Press, 1996), 327.

[34] Adrienne Rich, "Compulsory Heterosexuality and Lesbian Existence," *Signs* 5:4 (Summer 1980): 648.

genre of solitude, and speak with others, for others, using an ever-shifting plural pronoun, 'we'."[35] And Rich was careful about "we." She wrote in a journal entry that "someone writing a poem believes in, depends on, a delicate, vibrating range of difference, that an 'I' can become a 'we' without extinguishing others, that a partly common language exists to which strangers can bring their own heartbeat, memories, images."[36] Rich's Cold War poetry was already pointing her to this "delicate, vibrating range of difference" her late work would embrace, in part by creating not only opaque "I" speakers but also opaque "we" formations that can never include all women and continually critique the "I" who reaches for pluralities. Rich's constant self-scrutiny of the "I," the structures that create it, and its awareness of the other "I"s it cannot speak for is a staple of her work.[37]

In some ways, Bishop's and Rich's opaque lyric speakers extend, rather than repudiate, what confessionalism was already doing. But while Plath, for instance, uses extraterrestrial opacity to amplify the voice of her "I," Rich and Bishop do so to position their "I"s within larger collectives—never monolithic, always at least partly unknowable. Bishop's and Rich's poems, then, register but do not embrace either confessionalism or the critiques of the imperialism of lyric subjectivity begun by the Beats and extended through the Language movement, which outright rejected the lyric "I."[38] Like anti-lyric practitioners also working against confessionalism, Bishop worried about the "I"'s relationship to American imperialism and capitalism. Gillian White suggests that Bishop's poetics of "modesty" (as described by Lowell), anchored in her lyric "I," "can be regarded as a sharply *political* stance influenced by her mid-century cultural malaise," especially regarding the excesses of capitalism that shaped poetic language and made the very self into a commodity.[39] But Bishop could go only so far with that critique of subjectivity. Bishop and Rich liked the lyric "I"; it interested them intellectually, aesthetically, and

---

[35] Christopher Spaide, "'A Delicate, Vibrating Range of Difference': Adrienne Rich and the Postwar Lyric 'We,'" *College Literature* 47:1 (2020): 89–124, here 93.
[36] Rich, *What Is Found There: Notebooks on Poetry and Politics* (New York: W. W. Norton, 2003).
[37] In her introduction to Rich's *Collected Poems*, Rankine notes that Rich's scrutiny of the "I" led her to be "drawn to Rich's interest in what echoes past the silences in a life that wasn't necessarily my life." Claudia Rankine, Introduction to *Collected Poems: 1950–2012* (New York: Norton, 2016), xxx–xxxviii.
[38] For an expert framing of how Bishop's lyric "I" fits into mid-century debates about lyric, see Gillian White, *Lyric Shame: The "Lyric" Subject of Contemporary American Poetry* (Cambridge, MA: Harvard University Press, 2014), especially 42–50.
[39] See Gillian White, "Words in Air and 'Space' in Art: Bishop's Midcentury Critique of the United States," in *Elizabeth Bishop in the 21st Century: Reading the New Editions*, ed. Angus J. Cleghorn, Bathany Hicock, and Thomas J. Travisano (Charlottesville: University of Virginia Press, 2012), 255–73, here 256.

politically—especially, for Rich, as the personal was being championed as political in the 1960s.

But confessionalism didn't offer the kind of "I" that either sought. Part of Rich's career maps onto the breakthrough narrative: she rejected an impersonal "we" and claimed a personal "I" in her move from "cooked" to "raw" verse. But Spaide clarifies why reading Rich under the rubric of confessionalism so rarely works well: "even after liberating her 'I,' Rich never abandoned her collective 'we,' tirelessly testing new relations between singular and plural."[40] Alongside her strategically essentialist formulation of lesbian existence, Rich maintained an opacity in her lyric voices created out of an often inharmonious chorus of irreducible "I"s, many of whom are fundamentally unknowable. Bishop and Rich develop speakers who are not particularly like one another's but who nonetheless share a difficulty of classification: they are neither confessional nor absent.

In part, their distance from the confessional mode is an effect of neither Bishop nor Rich having full access to the kind of "I" articulated by Lowell, Bishop's closest literary friend. Writing to him to praise *Life Studies*, Bishop distinguished Lowell's genius from the limitations of women confessionals: "[Anne Sexton's] egocentricity...is simply that, and yours has been...made intensely *interesting*, and painfully applicable to every reader."[41] In an earlier letter predating his confessional turn, she had already noted her "envy" that Lowell's personal experience always came off as "significant, illustrative, American" (*WIA*, 247). Grobe spells out the obvious, that Bishop reads Lowell's presentation of self in this way because he's a straight white cisgender man: "Lowell's privilege allowed him to strike a delicate balance—to seem both private and public, both personal and social, but unique and representative."[42] Bishop lacked the identity coordinate to seem "public," "social," and "representative," and meanwhile she did not have the source material of domestic constraint and malaise that gave white women confessionals like Plath, Sexton, and even the early Rich entrance into the confessional mode. Bishop did not pursue a heterosexual nuclear family life, ignoring Lowell's awkward epistolary proposal to her that would have consummated their

---

[40] Spaide, "A Delicate, Vibrating Range of Difference," 117.
[41] Elizabeth Bishop, *Words in Air: The Complete Correspondence between Elizabeth Bishop and Robert Lowell*, ed. Thomas Travisano with Saskia Hamilton (New York: Farrar, Straus and Giroux, 2008), 327. Further references are noted in the text as *WIA*.
[42] Grobe, *The Art of Confession*, 38.

poetic friendship into a Cold War domesticity version of the "marriage plot."[43] Rich, meanwhile, recognized from the inception of the confessional movement that she would never have access to Lowell's universal "I." She recalls, "I had been taught that poetry should be 'universal,' which meant, of course, nonfemale."[44] She worked from the 1950s into the 1960s to pursue an "I" that registered the lesbian continuum.

Rich's and Bishop's poetry, unclassifiable by mid-century idioms, fits into what Thomas Travisano describes, in his critical veer away from the confessional narrative, as "a drive toward the self-exploratory."[45] Like other school-affiliated poets working at mid-century, both Rich and Bishop investigate the "I" against a complex backdrop of U.S. politics in a global context. But they thwart containment-era logic in a way their contemporaries did not. Their politics and aesthetic defy categorization as they turn to astronomy to articulate opaque queer lyric "I"s that are neither on display nor repudiated. This chapter will first take up Rich, as she fits more neatly into the breakthrough narrative, and will then consider Bishop's stranger contributions to an emerging queer lyric "I."

2.3 Adrienne Rich's Astronomical "We"

From 1953 to 1968, Rich seemed poised to make her breakthrough into confessional verse. In a series of three astronomical poems—"The Explorers," "Orion," and "Planetarium"—she moves from tight verse and an impersonal lyric "we" to free verse and a personal "I." Rich's early poem "The Explorers" (1953) conveys what Rich later identified as her early-career "absolutist approach to the universe."[46] It assumes a universal human perspective on planet Earth from the moon. It is, in fact, an Earthrise poem, written fifteen years before the famous NASA shot of Earth rising over the moon that I examine in Chapter 1, which led NASA-commissioned poet Archibald

---

[43] Hugh McIntosh argues that in their "conventional imaginary of domesticity and national devotion," Bishop and Lowell were "a queer couple." See Hugh McIntosh, "Conventions of Closeness: Realism and the Creative Friendship of Elizabeth Bishop and Robert Lowell," *PMLA* 127:2 (Mar. 2012): 231.

[44] Adrienne Rich, "When We Dead Awaken: Writing as Re-Vision," *College English* 34:1 (Oct. 1972), 18–30, here 44. Further references are noted in the text as "When We Dead Awaken."

[45] Thomas Travisano, *Midcentury Quartet* (Charlottesville: University of Virginia Press, 2009), 66. Travisano argues that the idea of the confessional conceals similarities we might otherwise notice across mid-century poets; Travisano, *Midcentury Quartet*, 32–70.

[46] Adrienne Rich, "Poetry and Experience: Statement at a Poetry Reading (1964)," in *Essential Essays: Culture, Politics, and the Art of Poetry*, ed. Sandra M. Gilbert (New York: Norton, 2018).

MacLeish to imagine a united human species living as "brothers" on planet Earth.[47] Rich turns a species-spanning masculine "we" gaze from the dead moon to the living Earth:

> Beside the Mare Crisium, that sea
> Where water never was, sit down with me,
> And let us talk of Earth, where long ago
> We drank the air and saw the rivers flow
> Like comets through the green estates of man,
> And fruit the colour of Aldebaran
> Weighted the curving boughs. The routes of stars
> Was our diversion, and the fate of Mars
> Our grave concern; we stared throughout the night
> On these uncolonized demesnes of light.[48]

The heroic couplets, with their origins in Greek heroic verse and their codification in eighteenth-century English didactic verse, evoke masculine adventures into unconquered realms of territories and knowledge. Like Odysseus, this "I" comes to long for the home it has left. Still positioning itself as a "we," the impersonal, even epically pitched, "I" describes nostalgia for Earth, as the speaker laments being "Across that outer desert from my home!" This style won Rich the 1950 Yale Younger Poets Award for *A Change of World* (1950), judged that year by Auden, whom in fact she sounds like in this poem. Auden was a self-fashioned poet of "human pluralities" whose masculinity allowed him, as Bonnie Costello has described, "to harness the rhetorical power of the first-person plural to posit and promote community, often where there is social fragmentation."[49] Rich's poem adopts Auden's community-invoking "we." Rich, though, would come to associate her 1950s style with domestic constraints on women, changing her lyric "we" into a lyric "I" that imagined a new kind of "we." In her essay "When We Dead Awaken," Rich describes her early formalism as "asbestos gloves, allow[ing] me to handle materials I couldn't pick up barehanded" ("When We Dead Awaken," 22). She then describes her move away from formalism, a system she notes "was formed

---

[47] Archibald MacLeish, "Riders on the Earth," *New York Times* (25 Dec. 1968), https://timesmachine.nytimes.com/timesmachine/1968/12/25/issue.html.
[48] Adrienne Rich, *Collected Poems: 1950–2012* (New York: Norton, 2016), 63. Further references are noted in the text as *CP*.
[49] Bonnie Costello, *The Plural of Us: Poetry and Community in Auden and Others* (Princeton, NJ: Princeton University Press, 2017), 3.

first by male poets," through poems of domestic confinement into cosmic escape, a move corresponding to her use of more open verse forms.

She would begin to use these forms in the early 1960s. Rich's sequence "Snapshots of a Daughter-in-Law" (1963) is often considered her "breakthrough" work.[50] It marks her first attempt to "write...directly about experiencing myself as a woman." While the poem is "in a longer, looser mode than I'd ever trusted myself with before," its neat quatrains evoke the claustrophobia of four walls of a domestic room. These close around the speaker, who in Section 2 looks outside, skyward, from her kitchen, where the speaker is not an "I" but a "she," as Rich later reflected, abstracted from herself ("When We Dead Awaken," 22). The jarring sounds of domestic labor open into the music of the celestial spheres: "Banging the coffee-pot into the sink / she hears the angels chiding, and looks out / past the raked gardens to the sloppy sky." The poem, a collection of "snapshots," emphasizes the acts of surveilling and being surveilled. The "sloppy sky" she "looks out" at transforms into a screen that does not watch her watch it: "Soon the night will be an eyeless quarry" (*CP*, 115). As Nelson writes of this passage, there is something comforting about being surveilled in one's own house, as it "relieves you of the terror of anonymity." The starless sky, in contrast, cannot reflect the self back to it and thereby produce an "I"; without surveillance, "you disappear into the black hole of the night."[51] But in "Orion" Rich finds herself reflected in the night sky, and in "Planetarium" she even becomes an astronomical formation.

Rich brought cosmic scales into her poems as she was writing poems of domestic constraint. In fact, Rich discusses *Snapshots* in the same essay where she describes her breakthrough through the vast cosmic horizons of poems like "Orion" and "Planetarium" (1968). Rich explains that she wrote "Orion" five years after *Snapshots* as she moved towards asserting the pronoun "I," breaking out of the house and into cosmic space. The poem's form blends traditional and experimental elements, written in seven isometric unrhymed sestets, three of which are enjambed and suggest less confinement than the snapped-shut stanzas of the earlier poem. The speaker apostrophizes the constellation Orion, channeling energy for her "I" through an egotism she at the time identified as masculine. Giving the constellation voice through the power of poetry, the speaker harnesses its energy as her own. Rich reflects that "Orion" was "a poem of reconnection with a part of myself I had felt I was

[50] See, for instance, James Longenbach, *Modern Poetry After Modernism* (Oxford: Oxford University Press, 1997), who numbers Rich among poets who had Lowell's "breakthrough back into life"; Longenbach, *Modern Poetry After Modernism*, 5.
[51] Nelson, *Pursuing Privacy*, 109–10.

losing—the active principle, the energetic imagination, the 'half-brother' whom I projected, as I had for many years, into the constellation Orion" ("When We Dead Awaken," 24). The concluding lines evoke both strength and confinement, as she imagines herself "out here in the cold with you / you with your back to the wall," ending the poem with its back against the wall of its own verse form (*CP*, 231).

Rich describes "Planetarium" as the "companion poem" of "Orion," "written three years later, in which at last the woman in the poem and the woman writing the poem become the same person" ("When We Dead Awaken," 25). In the 1960s U.S., planetarium visitors were supposed to feel a strong sense of "I." Rich wrote the poem during a proliferation of planetariums and planetarium programming in major U.S. cities in the 1960s designed to educate Americans of all ages in the space sciences and, thereby, in their proper place in the world. Planetariums have served as conduits between astronomers and the public since the 1930s in the U.S., but in the 1960s they became a vital source of citizen formation and national pride. Following the launch of Sputnik 1, a panicked U.S. government passed the National Defense Education Act of 1958, pumping over one billion dollars into STEM education, including space science, and prompting the building of hundreds of planetariums in schools across the nation. The planetarium became a place where children would be moved to become astronauts when they grew up and where adults would be moved to a childhood state of wonder.[52] With their pairing of stunning visuals, spooky music, lyrical narrations, the solitude of the dark, and comfortable chairs arranged in a communal sphere around a central projector, planetariums provide a theatrical experience of space that is at once private and communal. In its claiming of an "I" among "others," Rich's "Planetarium" challenges patriarchal claims to the universe while also evincing the success of a public educational program designed to put a strong sense of both American individualism and community at the core of space exploration.

Rich wrote "Planetarium" after visiting a planetarium with an exhibit on Caroline Herschel, "the astronomer, who worked with her brother William, but whose name remained obscure, as his did not" ("When We Dead Awaken," 25).[53] Rich uses the Herschels to respond to heteromasculine imperial

---

[52] Jordan D. Marché, *Theatres of Time and Space: American Planetaria, 1930–1970* (New Brunswick, NJ: Rutgers University Press, 2005), 4. Marché traces "the rise of American planetaria as a social phenomenon" (x). He devotes a chapter to the space age planetarium, "*Sputnik* and Federal Aid to Education," 119–36, that elaborates on the pedagogical principles behind the "space science classroom," meant to train students from a young age in scientific principles while inspiring wonder.

[53] Caroline Herschel was the first woman astronomer who earned a salary for the work. She discovered a galaxy and multiple comets between 1783 and 1797, and her update to star catalogues earned her accolades from the Royal Astronomical Society. Caroline Herschel's traditionally domestic

narratives that dehumanized women—even as she channels ways that Herschel has often been celebrated by American intellectuals. William Herschel, like the planetarium, was an important figure in American national identity. Herschel's intellectual conquest of the cosmos made him a popular figure of shapers of American culture from the Federalist Era through the Civil War. For John Adams, "A prospect into futurity in America is like contemplating the heavens through the telescopes of Herschel."[54] While shaping the identity of the "American scholar," Emerson would invoke Herschel in a different way, this time as a foil for the American soul-searching intellectual. Unlike astronomers "Flamsteed and Herschel," who, "in their glazed observatories, may catalogue the stars with the praise of all men," the American scholar labors "in his private observatory, cataloguing obscure and nebulous stars of the human mind."[55] Rich follows Emerson here in turning the heavens inward to magnify a lyric "I." At the same time, her poem is critical of American identity and power: her interest in cosmic disappearance in the poem registers ways in which "America" erases its most vulnerable citizens. As she put it in "What Kind of Times Are These": "this isn't a Russian poem, this is not somewhere else but here, / our country moving closer to its own truth and dread, / its own ways of making people disappear" (*CP*, 755). "Planetarium" tries to recover some of these disappearances through routes not available to the historically patriarchal science of astronomy.

While Emerson and Whitman looked to William Herschel as a foil for the poet's inner world, Rich looks to Caroline Herschel "and others" to articulate an "I" in relation to other women who are never entirely knowable to that "I." The epigraph restores what it can of women's histories while marking what remains irretrievable: "Thinking of Caroline Herschel (1750–1848), astronomer, / sister of William; and others" (*CP*, 301). The "and others" of the epigraph recalls how many women astronomers' contributions have been lost to the historical record, an analogue for the many women creators whose work was similarly lost. "Others" also records the othering of women, made into

---

labor that Rich describes—including keeping her brother's discovery logs and polishing the lenses of his telescopes—propelled William Herschel's more famous career. William Herschel (1738–1822) discovered in 1774 that the mysterious "nebulae" that Kant had earlier hypothesized might be "island universes" are clusters of stars and stellar nurseries. His catalogues of these objects (one in 1802 that recorded 2,500, and another in 1820 that added 5,000) revolutionized the study of astronomy—and were mostly compiled, exactingly and without credit, by Caroline Herschel, who also polished the lenses of his telescopes to maximize the light they could capture. For more on the Herschels' lives and work, see Michael Hoskin, *Discoverers of the Universe: William and Caroline Herschel* (Princeton, NJ: Princeton University Press, 2011).

[54] John Adams, *The Works of John Adams, Second President of the United States, Vol. 10*, ed. Charles Francis Adams (Boston: Little Brown, 1856), 6.

[55] Ralph Waldo Emerson, "The American Scholar," 31 Aug. 1837, http://digitalemerson.wsulibs.wsu.edu/exhibits/show/text/the-american-scholar.

the "monsters" of the night sky by the patriarchal histories of mythology and astronomy:

> A woman in the shape of a monster
> a monster in the shape of a woman
> the skies are full of them
>
> a woman   'in the snow
> among the Clocks and instruments
> or measuring the ground with poles'
>
> in her 98 years to discover
> 8 comets
>
> ...
>
> Galaxies of women, there
> doing penance for impetuousness
> ribs chilled
> in those spaces   of the mind
>
> (*CP*, 301)

The intralinear spacing evokes both constellations and the archival voids in women's histories, particularly their contributions to science. "Planetarium" has a documentary impulse to counter these historical erasures, quoting from Caroline Herschel's own journals about her life and work, but also documenting in blank spaces what cannot be retrieved. Like the Herschels' star catalogues, Rich creates her own verse catalogues of women's acts of discovery. She also undertakes this project in "Diving into the Wreck" and "Power," her homage to Marie Curie written in the same approximate form as "Planetarium." The gaps in the lines of the Curie poem signal historical erasures and the self-denials of a scientist whose "wounds came   from the same source as her power" (*CP*, 443). Herschel, Curie, and Rich herself all belong to "Galaxies of women" punished for their creations and discoveries. Another nameless "and other" Rich references is Northern Irish astrophysicist Jocelyn Bell Burnell. In 1967, Burnell discovered the first radio pulsars with her Ph.D. advisor. That advisor, Antony Hewish, won the 1974 Nobel Prize in Physics for this work; Burnell was not credited. Rich's poem, in the anonymity of "and others," foresees this elision.

Rich comes into an "I" that, like the history of women and their contributions to science, is riddled with unknowns. Rich moves from describing general "galaxies of women" to the particular woman of her speaker, who has the final stanza to herself:

> I have been standing all my life in the
> direct path of a battery of signals
> the most accurately transmitted most
> untranslatable language in the universe
> I am a galactic cloud so deep   so invo-
> luted that a light wave could take 15
> years to travel through me   And has
> taken   I am an instrument in the shape
> of a woman trying to translate pulsations
> into images   for the relief of the body
> and the reconstruction of the mind.
>
> <div align="right">(<i>CP</i>, 302)</div>

The passage describes the radio pulsars that Burnell and Hewitt discovered in 1967 through a process that the poem's form records. Using radio telescopes—which record radio signals rather than light visible to humans—they discovered pulsing emissions separated by 1.33 seconds that kept sidereal time (a timekeeping method based on measuring Earth's rotational rate against fixed stars). Initially, Burnell and Hewitt wondered if they had encountered a transmission from an advanced extraterrestrial civilization and nicknamed their project "LGM" for "little green men." Their theory, then, invokes both communication with radical others and the limits of the visual.

Rich's opening the poem with monsters in the night sky that also form a dyad of I–thou, watcher and watched, records this otherworldly relational context, in which women's bodies as seen by men become alien to themselves. The rhythmic disruptions created by spacing in the lines also records the temporal gaps between these pulsating emissions. The poet's body—and the poem itself—become "an instrument" that capture emissions that are not visible—a corrective metaphor in a poem about the politics of the visual, which inflected even the naming of constellations in the ancient mapping of women's bodies onto the night sky—and the ongoing naming of new celestial bodies after women in classical mythology. The poet compares herself to a dense celestial object to emphasize her own opacity, keeping her body from being neatly constellated into bright, observable stars. She is a "galactic cloud" that slows down the passage of light, "so invo-luted" that the word "involuted" does not even fix itself to a single line. The speaker is dispersed in both mind and body, not a bounded subject but instead one willing to be dispersed and reinvented to register the many lost experiences of women like and unlike herself. Instead of *being* seen, mapped into the night sky by the "neutral" male astronomer's gaze as figures for the mystifying and the unknown, Rich uses the observer effect to incite women to see and thereby change these systems: "What we see, we see / and seeing is changing."

In "Snapshots of a Daughter-in-Law," Rich's abstract "she" speaker watches a night sky utterly indifferent to her; in "Orion," the "I" identifies with the "masculine" energy of the constellation; by "Planetarium," the "I" who is, for the first time, "the woman writing the poem" experiences an embodied but non-visual relationship with the cosmos and with other women whose lives are not entirely lost to her, even if she cannot know them. "Planetarium" paves the way for Rich's later, more inclusive "we" that registers differences and respect for unknowability. When *Diving into the Wreck* won the National Book Award in 1974, Rich accepted it with co-nominees Audre Lorde and Alice Walker. Their co-written statement declared,

> We dedicate this occasion to the struggle for self-determination of all women, of every color, identification, or derived class: the poet, the housewife, the lesbian, the mathematician, the mother, the dishwasher…the women who will understand what we are doing here and those who will not understand yet; the silent women whose voices have been denied us, the articulate women who have given us strength to do our work. (*CP*, xviv–xlv).

These "silent women" appear in "Planetarium," but not just as voided spaces and lives on the page. Instead, the poem offers a mode of listening through an embodied lesbian continuum that exists outside of the male astronomical gaze. These women communicate through opaque radio signals, as opposed to brilliant, clear images delivered via the polished lenses of masculinity.

Rich's astronomy work has become part of a more woman-centric popularization of astronomy. Jocelyn Bell Burnell is not only a subject of "Planetarium" but also an advocate of using poetry in astronomy education. She co-edited the anthology *Dark Matter* (2008) of poems about astronomical phenomena, where she includes "Orion."[56] She also frequently reads poems in her public lectures, having become a cataloguer of poems about astronomical phenomena. In particular, she popularizes astronomy through poetry to stir women's interest in her profession: "To draw others, especially women, into science, I would like to give fair space to the human side of science," as she writes in her essay "Astronomy and Poetry."[57] She taps here into the gendering of genre; the

---

[56] Maurice Riordan and Jocelyn Bell Burnell, *Dark Matter: Poems of Space* (London: Calouste Gulbenkian Foundation, 2008).

[57] Jocelyn Bell Burnell, "Astronomy and Poetry," in *Contemporary Poetry and Contemporary Science*, ed. Robert Crawford (Oxford: Oxford University Press, 2006), 125.

"feminine" poem attracts "female" responses. Burnell reports that her aiming of poetry at women audience members is successful: "consistently, female audience members will come up to me afterwards and speak appreciatively of their inclusion."[58] This statement, however, rehearses a tired divide between poems as "feminine" and science as "masculine," one that Rich's own poetry rejects. Rich, in fact, used Burnell's pulsars to claim astronomy as a feminist pursuit. Pulsars, felt as embodied experiences for Rich's speaker, are invisible at the spectrum of light available to the human eye. They suggest, ultimately, what cannot be seen or recovered in patriarchal systems that privilege "sight" as a mode of knowledge and master women by reducing them to pretty or monstrous pictures. Rich's influence continues to be felt in astronomy popularization. In 2017, astrophysicist Janna Levin read Rich's "Planetarium" at the annual "Verse in Universe" gathering in New York City, an event that raises funds for climate science research through lectures and readings on the intersection of astronomy and poetry.[59] Rich's political and activist poetry has reverberated in public presentations of astronomy that increasingly challenge the field's patriarchal history.

2.4 Elizabeth Bishop's Queer Astronomy

The queer turn in Bishop studies was inaugurated by none other than Adrienne Rich. Rich's aptly named essay "The Eye of the Outsider" implicitly links lesbian poetics to opacity. Taking stock of Bishop's impact on her own work shortly after her "Compulsory Heterosexuality" essay, Rich recalls,

> I had felt drawn, but also repelled, by Bishop's early work—I mean *repel* in the sense of refusing access, seeming to push away.... in part they were difficulties I brought with me, as a still younger woman poet already beginning to question sexual identity, looking for a female genealogy, still not yet consciously lesbian. I had not then connected the themes of outsiderhood and marginality in her work, as well as its encodings and obscurities, with a lesbian identity. I was looking for a clear female tradition; the tradition I was discovering was diffuse, elusive, often cryptic.[60]

---

[58] Burnell, "Astronomy and Poetry," 126.
[59] Maria Popova, "Planetarium: Astrophysicist Janna Levin Reads Adrienne Rich's Tribute to Trailblazing Women in Science," *BrainPickings*, 27 Apr. 2017, https://www.brainpickings.org/2017/04/27/janna-levin-reads-planetarium-by-adrienne-rich/.
[60] Adrienne Rich, "The Eye of the Outsider: The Poetry of Elizabeth Bishop," Boston Review 8:2 (1983): 15–17, here 15.

The lineage Rich identifies, like the poetry itself, is encoded and obscure. The "elusive" and "cryptic" dimensions Rich finds in Bishop's work speak to the Cold War surveillance culture in which Bishop wrote subversively queer poetry.

Rich's essay registers Bishop's interest in optical science, which extended to her fascination with astronomy.[61] In fact, few postwar poets have had as abiding an interest in astronomy as Bishop. As her friend Wheaton Galentine reflected, "She had a lot of scientific interests you wouldn't necessarily associate with a literary person. She was interested in astronomy and used to talk about space flight."[62] Indeed, Bishop's career exactly spans the development of air and space technologies that led humans first off of Earth and then back to it. One of her earliest drafts, "Good-Bye," takes place in an airport terminal; it was written during the first U.S. commercial aviation boom between 1931 and 1934, a period in which the number of Americans to fly annually rose from 6,000 to almost half a million. Her class privilege gave her access to the aerial views that in that period were the exclusive domain of male pilots. Aerial views and the problems with them would preoccupy her space age career. Her first published poem, "The Map," takes a bird's-eye view of Nova Scotia's coastline. Bishop's mid-career poems "The Shampoo," "Insomnia," and "The Armadillo" turn their gaze from Earth's surface as seen from the sky to deep space as seen from the Earth. *Geography III*, published a few years after the final moon landing, suddenly turns *back* to planet Earth. Its cover pictures the terrestrial globe with cartographers' instruments of measurement; the epigraph asks, "What is the Earth?" (*G*, 1). The simple question, taken from a nineteenth-century geography primer, captures the cultural difficulty of reading planet Earth as it emerged as a new kind of figure in the late 1960s: an extraterrestrial object, an otherworld—and for Bishop, an implicitly queer one. She queers the planet in this late work as she writes both within and against American imperial rhetoric.

This snapshot of Bishop's career of the air and space age underscores what a tricky and hard-to-classify figure she is; her preference for traditional lyric coupled with the sometimes classist and imperialist dimensions of the work complicate what scholars increasingly identify as a progressive politics. The

---

[61] Bishop read Newton's *Opticks* and worked, for about a day, in an optical shop in Key West before the Second World War; she wrote to Marianne Moore about admiring both the optical instruments and "the theory of the thing," of "why the prisms go this way or that way, or what 'collimate' and 'optical center' really mean"; see Elizabeth Bishop, *One Art: The Selected Letters*, ed. Robert Giroux (London: Pimlico, 1994), 116.

[62] Elizabeth Bishop, quoted in Fountain and Brazeau, *Remembering Elizabeth Bishop*, 381.

scientific turn in Bishop studies has tied her many scientific interests, particularly Darwin, to a surprisingly subversive politics and aesthetics. Bishop called Darwin her favorite writer and "one of the people I like best in the world."[63] Her oft-quoted Darwin letter expresses her admiration for his scientific-poetic method of opacity:

> Reading Darwin, one admires the beautiful solid case being built up out of his endless heroic *observations*, almost unconscious or automatic—and then comes a sudden relaxation, a forgetful phrase, and one *feels* the strangeness of his undertaking, sees the lonely young man, his eyes fixed on facts and minute details, sinking or sliding giddily off into the unknown. What one seems to want in art, in experiencing it, is the same thing that is necessary for its creation, a self-forgetful, perfectly useless concentration.[64]

Bishop's description of Darwin emphasizes how looking renders the self strange. In this way, Bishop's reading of Darwin has bearings on her queer and ecological interests, linked through her fascination with epistemological limits and the effects of opaque knowledge on subjectivity and self-knowledge. Darwin also offered Bishop a way to conceptualize homosexuality in natural terms, in contrast to sexology's model of the "invert" and the homosexual as an "abomination against nature." Susan McCabe argues that "Darwin's vivid presentation of slow transitions over extended geological time, peculiar forms, gradual adaptations, and transitional states" and his focus on "nonreproductive sexuality" informed Bishop's "sense of sexual and aesthetic deviation."[65] Sarah Giragosian brings together Bishop's queer and ecological interests to argue, compellingly, that Bishop was drawn to "Darwinian evolution as a relational and creative force with a queer political potential," and that Bishop draws from the permeability between human and nonhuman animals described in Darwinian evolution to "articulate a posthuman queer subjectivity," in which it is possible to imagine noncoherent forms of selfhood that exist outside of nationally scripted linear temporalities.[66] Bishop's critical

---

[63] Brett Millier, *Elizabeth Bishop: Life and the Memory of It* (Berkeley: University of California Press, 1993), 356; Bishop, *One Art: The Selected Letters*, 543. For a definitive account of Bishop's interest in Darwin, see Jonathan Ellis, "Reading Bishop Reading Darwin," in *Science in Modern Poetry*, ed. John Holmes (Liverpool, UK: Liverpool University Press, 2012), 181–93.

[64] Letter to Anne Stevenson, 8 Jan. 1964, in Elizabeth Bishop, *Poems, Prose, and Letters* (New York: Library of America, 2008), 861.

[65] Susan McCabe, "Survival of the Queerly Fit: Darwin, Marianne Moore, and Elizabeth Bishop," *Twentieth-Century Literature* 55:4 (Winter 2009): 547–571, here 549.

[66] Sarah Giragosian, "Elizabeth Bishop's Evolutionary Poetics," *Interdisciplinary Literary Studies* 18:4 (2016): 475–500, here 477.

legacy, then, is uncategorizable. She is at once remembered as a "pleasantly idiosyncratic maiden aunt" and a "minor poet" of the miniature, as a review of her work in the Apollo 11 moon landing issue of *Life* magazine describes her.[67] At the same time, she is increasingly appreciated for her queer politics and aesthetics. Her astronomy cuts across those divided legacies.

Bishop turned her own fascination with optics against the Cold War surveillance state. She wrote only one poem during her miserable year as poetry consultant at the Library of Congress from 1949 to 1950, a year of peak paranoia about treasonous homosexuals. "View of the Capitol from the Library of Congress" stares at the building symbolizing the government that was actively surveilling queer people in 1949. Instead of it watching her, Bishop's speaker watches it. Camille Roman and Stephen Axelrod both identify this poem as an implicitly queer critique of Cold War America. Roman argues that "Bishop's attempts to resist the invasiveness of patriotic courting as a cultural producer while appearing to accommodate it reveals a rhetorical and real strategy for dealing with Cold War cultural surveillance."[68] The containment context of the poem is underscored by its original publication venue, in the *New Yorker* of 7 July 1951 beneath a cartoon of a well-dressed woman, legs demurely crossed, visiting her husband in prison. The caption reads, "How do you expect the children to respect you if you don't get time off for good behavior?"[69] The cartoon comes out of a cultural moment in which women's conventional behavior within a heterosexual marriage was imagined to be a moral guard against the spread of communism; its prison setting also evokes the surveillance state. Axelrod extends Roman's insights to account for specific formal decisions that evoke Bishop's development of a "queer politics." Axelrod argues, "the poem speaks in opaque and even cryptographic figures that allow it to escape Cold War surveillance and censorship, yet this oblique style contains distinctly insubordinate codes."[70] One of these codes is its many instances of visual and sonic inversions, picking up on sexology's model of the homosexual as an "invert." Indeed, the entire conceit of the poem is the government-employed lesbian poet overlooking the Capitol, the synecdoche for the government that would have been very interested in watching her.

---

[67] Charles Elliott, "Elizabeth Bishop: Minor Poet with Major Fund of Love," *Life Special Issue: Off to the Moon* (4 July 1969), 13.

[68] Camille Roman, *Elizabeth Bishop's World War II–Cold War View* (New York: Palgrave Macmillan, 2001), 131.

[69] Syd Hoff, "How do you expect the children to respect you if you don't get time off for good behavior?," *New Yorker* (7 July 1951), http://www.sydhoff.org/pages/TheNewYorker.html.

[70] Steven Gould Axelrod, "Elizabeth Bishop and Containment Policy," *American Literature* 75:4 (Dec. 2003): 849–50.

This is not to say that Bishop was being watched by the government in an official capacity. But this, as Roman suggests, was in part the result of what Merrill described as Bishop's "instinctive, modest, life-long impersonation of an ordinary woman."[71] Nonetheless, communist witch-hunts hit close to home for Bishop during her consultancy. Trying to escape the life of a government employee, she spent a lot of the year away at the Yaddo arts colony in Saratoga Springs, New York, where in 1949 Lowell—who had recommended Bishop for the poetry consultancy he had recently held himself—accused its director, Elizabeth Ames, of being a communist. Lowell accused Ames in response to her inviting pro-communist writer Agnes Smedley to the colony. The homosexual dimension of containment fears ripples under the surface of Lowell's allegations: Ames and Smedley, like Bishop, were unmarried, and Smedley's "lesbian tendencies" were well known.[72] Whether in a government office or at a private arts colony, containment culture infiltrated Bishop's creative space. Directly after what she called "that dismal year in Washington," Bishop moved to Brazil for sixteen years, where she lived with her partner, Brazilian architect Lota de Macedo Soares (*WIA*, 143). Merrill, again using an optical metaphor, would write that he and Bishop preferred "the camouflage of another culture."[73]

Bishop frequently uses tropes of astronomical inversion to evoke queerness. Timothy Materer notes that both Merrill and Bishop "play on the word *inverted* to denote the view of homosexuality as abnormal and yet also to suggest seeing the world in a fresh way."[74] Even in her earliest work, images of astronomical inversion abound. In "The Man-Moth," the eponymous hero of the poem, often read as a queer figure, "makes an inverted pin, the point magnetized to the moon."[75] Elsewhere, Bishop describes a lonely night in a room watching the inverted flight paths of five flies, comparing them to planets in retrograde motion: "Like five planets, only they both went back and forwards: / on their track / They let themselves drift back."[76] Her astronomical love poems abound with queer inversions. In "Insomnia," the speaker, kept awake by the loss of her lover, watches the Sapphic moon's reflection reversed in the

---

[71] James Merrill, "Elizabeth Bishop (1911–1979)," in *Elizabeth Bishop and Her Art*, ed. Lloyd Schwartz and Sybil P. Estess (Ann Arbor: University of Michigan Press, 1983), 259–62.

[72] See Ruth Price, *The Lives of Agnes Smedley* (Oxford: Oxford University Press, 2005), especially 158.

[73] James Merrill, *A Different Person: A Memoir* (San Francisco, CA: Harper, 1994), 141–2.

[74] Timothy Materer, "Mirrored Lives: Elizabeth Bishop and James Merrill," *Twentieth Century Literature* 51:2 (Summer 2005): 179–209.

[75] Elizabeth Bishop, "The Man-Moth," in *Poems* (New York: Farrar, Straus and Giroux, 2011), 16. Further references are noted in the text as *P*.

[76] Draft of "In a Room," dated "Seville 1936," in Elizabeth Bishop, *Poems*, 269.

mirror and longs for "that world inverted," "where the heavens are shallow as the sea / is now deep, and you love me" (*P*, 68). Astronomical estrangement, and the "unnatural" reflection of the Sapphic moon in the mirror, intensify the mood of unrequited desire. In "The Shampoo," Bishop invokes astronomical deep time that eclipses the fleeting human lifespan to make an inverted *carpe diem* argument. Rather than urging her addressee to speed up their courtship because time is flying, she urges her "pragmatic and precipitate" lover to embrace the slowness of cosmic time and enjoy the leisurely intimate banality of having her hair washed: "—Come, let me wash it in this big tin basin, / battered and shiny like the moon" (*P*, 82). The love poem honors the aging female body, rendering the process of aging natural by comparing them to the moon's life cycle.

Another of Bishop's Brazil poems, "The Armadillo" (1957), looks out at the fully "inverted" night sky of the southern celestial hemisphere relative to the northern. The speaker watches the illegal fire balloons launched for the St. John's Day carnival in Rio rise into the night, where they meet a series of visually similar celestial objects. The voice is at once scientifically precise—concerned with measurement and the naturalist's work of telling things apart—and constantly revising its own conclusions about its perceptions. Mid-line revisions and qualifications—"that is," "or"—create a slippery geographical and subject position augmented to the scale of cosmic unfathomability. The poem is interested in what cannot be known: the speaker finds it "hard / to tell [the balloons] apart from stars," which she then corrects to "planets," only to specify that she means "tinted ones" that may be either "Venus going down, or Mars," or "the pale green one" that she makes no attempt to name (*P*, 101). The rhyme of "stars" and "Mars" emphasizes the unreliability of perception, with rhyme artificially linking two unlike celestial bodies. The "Southern Cross" constellation introduces yet another potential corrective to identifying these celestial objects: the supposed planets may, in fact, be stars after all, since the Southern Cross contains the "Jewel Box cluster" of bright, colorful stars named by John Herschel (Caroline Herschel's nephew); this is consistent with "tinted" or "green" objects the poem describes. The Southern Cross is used for navigation, much as Polaris is in the northern hemisphere. The poem, then, rehearses its inability to assert knowledge of what it sees and, by extension, of where it is in time and space.

"The Armadillo" was published in the *New Yorker* six months before the launch of Sputnik, which inverted the cosmic gaze. Geographer Peter Sloterdijk argues that Sputnik inaugurated an "inverted astronomy":

something has got under way and led to reversal of the oldest human gaze: the first satellite was put in orbit above the earth. Soon afterwards the area of space close to the earth was teeming with satellite eyes, which provide technical implementation of the ancient phantasm of God looking down from high in the heavens. Ever since the early sixties an inverted astronomy has thus come into being, looking down from space on the earth rather than from the ground up into the skies.[77]

This inverted look back at Earth, which rearranged cosmic scale, preoccupies *Geography III*, the book Bishop assembled and published following her return to the U.S. from Brazil.[78] The epigraph's query—"What is the Earth?"—highlights the shifting importance and meanings of the human on a planet that keeps changing size.

Geography III follows the culturally inverted gaze from deep space back to Earth, taken in at different scales. The book jacket displays an antique terrestrial globe surrounded by cartography instruments, evoking scale, measurement, and the world-making work of poems.[79] All of its poems take in Earth's surface at different scales and from different perspectives. "One Art" magnifies its speaker's losses by zooming out from "keys" to "houses" to "two cities" to "a continent," then back to the small scale of a singular "you" whose loss is, after all, a "disaster" (*G*, 40). "Objects and Apparitions" imagines the lonely view of "God all alone above an extinct world" (48). "Night City," with its abstracted perspective *"from a plane,"* takes in the features of a wartime city below, watched by the "blackened moon" that evokes the Apollo program (19–20).[80] "12 O'Clock News" also includes deadened lunar imagery from an

---

[77] Peter Sloterdijk, quoted in Wolfgang Sachs, *Planet Dialectics: Explorations in Environment and Development* (London: Zed Books, 1999), 111.

[78] These poems are hard to date because Bishop revised so extensively across her career. Some, like "The Moose" and "12 O'Clock News," were begun in the 1940s. But most of the poems were written or substantially revised during the 1960s.

[79] Bishop had a lot of input on the book jacket and even designed an early version herself; she wrote to Anny Baumann, "I actually got out my watercolors and designed my own book jacket for...*Geography III*....[It] looks like an old-fashioned school-book, I hope"; Bishop to Baumann, 24 Dec. 1975, *One Art*, 602. This was also apparently Bishop's favorite dust jacket of all her books (*One Art*, xxi). The final book jacket, designed by Cynthia Krupat, fit Bishop's vision: "I discussed my idea of a book jacket [for *Geography III*] with [Bob Giroux of FSG], and said I'd written to Cynthia [Krupat] a few days earlier. I wish you could have heard how flattering he was about Cynthia's work—said of course she'd know how to make the jacket I have in mind if anyone did, and what good work she does, etc. etc." (Bishop, *One Art*, 605).

[80] Marit MacArthur discusses the specter of the moon landing in "Night City," noting that "here a hostile, airless lunar atmosphere comes down to Earth, extinguishing all Romantic associations with the 'blackened moon'" (272). Marit MacArthur, "One World? The Poetics of Passenger Flight and the Perception of the Global," *PMLA* 127:2 (Mar. 2012): 264–82.

aerial view of enemy terrain, with a newscaster who takes on the bird's-eye view of a combat pilot to describe the objects on her writing desk as a war zone. In "Crusoe in England," Crusoe attempts an aerial view of his topography but feels like a "giant" on his short volcanoes with their "heads blown off" by eruptions, and can see only the horizontal, sea-level view of "islands spawning islands" (9, 16). The volume's final poem shrinks these aerial scales to their most local form, ending only "Five Flights Up" in a New York City apartment, giving the speaker just enough height to see her neighbor's dog "rush around in the fallen leaves" (50). The volume's materiality, moreover, is itself an exercise in scalar distortions; its mere ten poems and slender spine contrast with the globe on the cover; the typeface, moreover, is huge, even as the text shrinks into wide margins.

The volume reminds us that scale is also a musical term. In "Crusoe in England," Crusoe recalls playing his homemade flute: "I think it had the weirdest scale on earth" (13). The poem records the weird scale of the self as it fails to connect with others, specifically Friday. The lonely Crusoe wishes he could "know enough of something," such as "astronomy," to take him out of the monotony of a constant lonely encounter with himself. The poem's "weirdest scale on Earth" is tied specifically to queer desire. In his loneliness and boredom, Crusoe watches Friday's "pretty body" and regrets that they cannot "reproduce," but most of the poem is not about Friday at all. The final two lines alone reveal the entire thing to have been an elegiac love poem: "—And Friday, my dear Friday, died of measles / seventeen years ago come March" (18). The poem demonstrates an inverted scalar relationship to desire: the final two lines carry the most weight of the lengthy poem. Both the imagery and the musicality of the volume manipulate scale as part of its overall effects of queer opacity.

Many critics approach *Geography III* for its queer content rather than its form. Lee Edelman famously suggests that the speaker of "In the Waiting Room" feels shy and horrified by her sexual attraction to the photographs of naked tribal women in the *National Geographic* she reads.[81] Most critics note Crusoe's attraction to Friday, and many connect his anguish at Friday's loss to Bishop's processing of Soares's suicide.[82] And "One Art" is dedicated to Alice Methfessel, Bishop's editor and partner, whom Bishop feared losing at the time of writing. However, the volume is at least as striking for its queer

---

[81] Lee Edelman, "The Geography of Gender: Elizabeth Bishop's 'In the Waiting Room'," *Contemporary Literature* 26:2 (Summer 1985): 179–96, here 194.

[82] See, for instance, McCabe, "Survival of the Queerly Fit," 566–7; and Millier, *Elizabeth Bishop*, 449.

content as for its underexplored queer form. Its salient queer formal elements include the "inverted" astronomical gaze that structures the entire volume as well as its lyrical strategies of opaque relations—specifically its mode of manipulating scale in both earthbound and extraterrestrial contexts. Scale, the volume reminds us, depends on the existence of a subject who measures phenomena against herself. To record and constantly recalibrate scale is to experience a human subject in flux, and to keep that speaker from being fully read and classified.

Geography III is self-fashioned as an advanced geography primer, based on Monteith's *First Lessons in Geography*. As the imagined third text in the National Geographical Series of U.S. public education textbooks, *Geography III* offers more advanced lessons than Monteith's. But it uses lyric tools to train readers in how to read opaquely, rather than for the epistemological mastery *First Lessons in Geography* advances. The textbook joined a nineteenth-century effort in Europe and the U.S. to educate schoolchildren in their dominant place in the world order. As Denis Cosgrove has it, the title of a popular school atlas, "*The English Imperial Atlas and Gazetteer of the World*...clearly expresses the principal educational message conveyed by its images."[83] An edition of *First Lessons in Geography* from 1856 all but renders planet Earth *as* the U.S.; preceding Lesson VI is an illustration of planet Earth rotating away from sun into night with only one country marked: "United States." *First Lessons in Geography*, originally published in the earliest years of American public schooling, underscores how geography—as the study of the Earth and its planetary neighbors—was part of the foundation of American public education. This nineteenth-century pedagogy of cosmic mastery was revived during the space race. Bishop wrote and revised some of the poems in the wake of the National Defense Education Act (a likely source of funding for the exhibit behind Rich's "Planetarium"). Many of the poems of *Geography III* are pedagogical, including "Crusoe in England" (based on Defoe's didactic novel) and "One Art," which instructs readers in how to lose. "In the Waiting Room," the opening poem of the volume, is likewise pedagogical, thematically and formally concerned with relations and with reading. Bishop drafted the poem in the late 1960s as Rich was working on "Planetarium." Like "Planetarium," the poem seems to suggest an alignment of "the woman in the poem and the woman writing the poem"; the speaker is even named "Elizabeth," and it is based on Bishop's autobiographical essay about

---

[83] Cosgrove, *Apollo's Eye*, 225.

accompanying an aunt to a dental appointment.[84] Both "Planetarium" and "In the Waiting Room" explore coming into selfhood with a space age flair, articulating new relations between "I" and "we."

"In the Waiting Room" is an exercise in reading, and failing to read, others. Six-year-old "Elizabeth," apparently the child version of the grown-up poet, has accompanied her Aunt Consuelo to a dental appointment. While waiting for the procedure to end, "Elizabeth" sits in the waiting room reading an issue of *National Geographic* ("I could read," she emphasizes) in shy avoidance of the strangers in the room. After closing the magazine, she hears an "*oh!*" of pain that seems to come at once from her aunt and from her own mouth. She then feels the disorienting shock of being "an *I*" among others, a sensation she likens to falling off the planet into space. Then the shock settles, and she finds herself back in the waiting room, shut off from the First World War raging outside and far away. Like "Planetarium," "In the Waiting Room" is set in an enclosed room that the poem imaginatively expands into extraterrestrial space. The setting of a waiting room sealed off from a distant war that doesn't quite intrude has an analogue in the Cold War context in which the poem was written, another war fought ostensibly "outside" the borders of the nation, in locales as various as Southeast Asia, the Caribbean, and outer space. This Cold War poem scrutinizes borders that won't stay shut. The breakdown of distinct boundaries and borders in the poem, including those between self and other, speaks to a merging of "domestic" and "foreign" space brought together through the master Cold War metaphor of containment. The speaker, like the poem, is not a bounded "I"; she is an unbounded and opaque text. Moreover, the title of *National Geographic* links the poem back to the *National Geographical Series* that published *First Lessons in Geography*. But perhaps the biggest event of reading in the poem is the attempt to read the "*oh!*" of pain—not only where it comes from but also what it might signify. As Edelman describes, it is "a cipher,"; it is "a zero, a void," standing in for all the ways social structures render women unreadable.[85] This "*oh!*" of pain invites lyric reading methods in a poem that is narrative and even features the act of reading prose and pictures in a magazine. This quasi-apostrophic "*oh!*" replaces linear narration with lyric time and space. "In the Waiting Room" offers a lesson in how to read the world—specifically, it puts forth a mode of queer lyric reading.

---

[84] Elizabeth Bishop, "The Country Mouse," in *Complete Prose* (New York: Farrar, Straus and Giroux, 1984), 13–34.
[85] Edelman, "Geography of Gender," 196.

The materiality of the popular science magazine shapes both "Elizabeth"'s experience with reading and the reader's reading of the poem. *National Geographic* presents itself as eminently readable, with straightforward, catchy stories punctuated by captivating pictures. They are nineteenth-century geography primers for grown-ups, amounting to a form of public education; they're strategically placed in waiting rooms to fill otherwise wasted time with productive learning about one's world. Combining naturalist and anthropological stories with documentary photography, the magazine suggests to American readers (the audience inscribed in the magazine's name) that it offers them transparent access to the world they have the right and duty to understand as global citizens. "Elizabeth"'s voice often matches the documentary matter-of-factness of the prosaic tone of the magazine. She even begins with a place and ends with a date, framing the poem in its narrative coordinates: "In Worcester, Massachusetts" and "the fifth / of February, 1918" (3, 8). At first, she reads the *National Geographic* as American citizens are supposed to, productively filling the wasted time of waiting by undertaking a global education. After studying photographs of volcanoes, "Elizabeth" turns to the next story; she stares at the photographs taken by American adventurers Osa and Martin Johnson, who frequently photographed peoples of East and Central Africa, South Asia, and the South Pacific to display them for visual and epistemological mastery to "global" American citizens. While Bishop fudges the facts here—there was no issue of *National Geographic* in February or March of 1918 that featured stories on both volcanoes and Africa[86]—this pairing is demonstrative of how *National Geographic* relegates non-Western peoples to prehistory, aligning them with an early evolutionary phase in the planet's natural history.

As "Elizabeth" studies the photographs, the narrative time of the *National Geographic* breaks down and is replaced with a lyrical temporality of aesthetic contemplation, bringing them out of the deep past and into the speaker's present world. The magazine, and the poem, display the bodies of "black, naked women" for interpretation; they are aestheticized, lyricized, captured in a suspended moment in time for "Elizabeth" to peruse. When "Elizabeth" closes the magazine, a fully lyrical moment erupts, displacing narrative with the cry of apostrophe: "Suddenly, from inside, / came an *oh!* of pain" (4). This "*oh!*," like the "*O falling fire*" of "The Armadillo," invokes apostrophe without using it. It is an example of the "planetary apostrophe" I describe in Chapter 1

---

[86] For a discussion of the accuracy of Bishop's National Geographic contents, see Edelman, "The Geography of Gender," 183–4.

that travels with Earth photography, as poets attempt to read the planet or record failures to address it. Evoking what Jonathan Culler calls the "embarrassment" of the figure of apostrophe, "Elizabeth" is almost "embarrassed" "to overhear / a cry of pain" in her aunt's voice, a metalyric moment of a private utterance being overheard.[87] Her embarrassment, though, is stalled by the shock of recognizing the cry not as her aunt's, but as her own: "my voice, in my mouth" (5). The "*oh!*" confuses the I–thou poles of lyric address and brings distinct parts of the poem into indecipherable relation to one another. The quasi-apostrophe connects Elizabeth not just to her aunt but also to the rounded "inside of a volcano," the "breasts" in *National Geographic*, and even the O of planet Earth. The difficulty of reading that "*oh*" gives Elizabeth "the sensation of falling off / the round, turning world / into cold, blue-black space" (5–6).

This image is strikingly similar to NASA Apollo photographs of planet Earth suspended in an apparent void of space that began appearing in popular magazines in the late 1960s. In 1967, *National Geographic* (the very magazine that "Elizabeth" reads) included a "full color photograph of Earth" captioned with a line from a George Meredith sonnet about Satan looking down on the world from his position "Above the rolling ball in cloud part screened."[88] While Meredith's lines evoke the vertical sublime—"above the rolling ball" looking down on Earth—"In the Waiting Room" redirects this gaze horizontally. It subverts Cosgrove's vertical Apollonian gaze by looking *out*, relationally, rather than *down*, imperially. As she realizes she is "an I" among other "I"s, she tries to apprehend herself in horizontal relation to others, her descriptions remaining at the eye level of her speaker, noticing the knees, hands, and clothing of the strangers in the waiting room.[89] Her accretion of details about these strangers recalls the Darwin of Bishop's letter to Stevenson, who gathers observations that send him toward the unknown rather than mastery; Zachariah Pickard notes that "In the Waiting Room," with its lists of details and attention to slight variations between people, is a Darwinian poem.[90] As the "I" compounds observations, it, like the Darwin of Bishop's letter, confronts the strangeness of the Darwinian pursuit and "slides…off into the unknown." The lamps in the room, a worn metaphor for

---

[87] Bishop, *Geography III*, 5, 7. Jonathan Culler, *Theory of the Lyric* (Cambridge, MA: Harvard University Press, 2015), 190.

[88] Quoted in Kenneth F. Weaver "Historic Color Portrait of Earth from Space," *National Geographic* 132:5 (1967): 726.

[89] Bishop, *Geography III*, 6.

[90] Zachariah Pickard, *Elizabeth Bishop's Poetics of Description* (Montreal, CA: McGill-Queen's University Press, 2009), 68–70.

illumination and knowledge, are dim; the late afternoon winter room occludes the speaker's perceptions, making some details "shadowy." The poem remains at the eye level of the six-year-old *except* when she looks down at the stories in the magazine. "Elizabeth" does not "glance" horizontally at the bodies in the magazine but, rather, stares down at her own lap, where she holds them in shrunken form; these bodies are not modestly and dimly revealed by lamplight but, rather, become sources of illumination themselves, the women's necks compared to "necks of light bulbs" (4). The poem, like the magazine, uses this moment of visual transparency as an opportunity for enlightenment. Its white American speaker moves into a lyric subject position by turning these Black bodies into objects of radical alterity, others to be visually mastered and assimilated so that "Elizabeth" can feel herself as an "I." This mastering gaze is governed by the magazine's materiality: held in a lap, below eye level, it shrinks pictures—of humans, of volcanoes, of the planet—to a size that is digestible to human subjectivity.

But confronted with a shock of otherness, "Elizabeth" cannot digest what she sees, prompting her to wonder "What similarities—" "held us all together /or made us all just one?" (6–7). This passage evokes the space age propaganda the *New York Times* commissioned MacLeish to produce, where he wrote of a common brotherhood who were "riders on the Earth together...brothers on that bright loveliness in the eternal cold."[91] But "In the Waiting Room" offers no idealized, harmonized vision of human belonging. Writing of these Apollo photograph intertexts, Susannah Hollister notes, "The poem's publication in 1971 made it a timely comment on the era of space travel....For the child in the poem, unlike the promoters of the space program, the mental image of the earth brings no easy reassurance that she belongs to a unified whole."[92] The poem arrives at the horror of imagining a whole by the racist, colonialist logic promoted by *National Geographic*, which lays out images of exotic others and elsewheres against which its American readership can define itself.

But, as Jahan Ramazani argues by way of Spivak's planetarity, the poem's claiming of an "I" is not as simple as that: "what distinguishes the encounter with otherness...from the Johnsons' exoticist language and unselfconsciously triumphalist photographs is the poet's foregrounding the precarious act of self-fashioning in a differential relation to the cultural other." This mode of

---

[91] MacLeish, "Riders on the Earth."
[92] Susannah L. Hollister, "Elizabeth Bishop's Geographic Feeling," *Twentieth-Century Literature* 58:3 (Fall 2012): 399–438, here 425–6.

self-fashioning is, for Ramazani, a specifically lyrical act, as Bishop foregrounds "self-address and self-nomination" through meta-attention to lyric address. In this "quintessentially 'lyric' moment of emerging self-consciousness," Ramazani observes, "the vulnerable subject turns to address itself."[93] This horizontal method of self-address mirrors Spivak's distinction between the self who becomes a "global agent" through mastering a colonized other and a "planetary accident" which conceives of itself as one part of planetary processes "in a galactic and para-galactic alterity."[94] Bishop's "I," imagining itself within a larger cosmos, is just "an *I*" among others not connected by MacLeishian sameness; instead, the poem evokes "everything only connected by 'and' and 'and,'" linked precariously by a hard-to-read "*oh!*" (*P*, 58).

"In the Waiting Room" captures the artifice and opacity, indeed the queerness, of this relationality. The passage hinges on "similarities" rather than "sameness." Like many of Bishop's poems, this one teems with similes that hold open the space a metaphor would purport to close, that call attention to the differences and unknowable sites they mark. Even the apostrophe is only similar to itself, not O but "*oh!*," calling out but to nothing it can name, just as its subject cannot be located. The poem plays with the words "like" and "likeness," a kind of simile for a simile. The poem hinges not just on likeness but also on "unlikeliness," and even on unlikeness. "[H]ow '*unlikely*,'" she thinks, that they should all be there together, in the same room in 1918, linked by the suspending lyric time of waiting captured in the *oh* and the planet photographically fixed in space (*G*, 7). With its almost apostrophic *oh!* of pain, the poem shows us the opacity and the strangeness involved in imagining a human collective. Inverting the cosmic gaze back to Earth and picturing its relationality as an unreadable "*oh!*" of pain, the *oh* of poetic artifice invites us to read it, only to mark what we can and should never understand about ourselves or others.

Rich's and Bishop's queer astronomy is anchored in the postwar U.S. even as it registers Cold War reverberations from beyond the nation's borders. Chapter 3 turns to the career of Kashmiri American poet Agha Shahid Ali, who is frequently read as a postcolonial poet but seldom as a Cold War one. It opens up the queer dimensions of poetic form and astronomy in a more complex transnational context by focusing on an object of perennial poetic and American fascination: the moon.

---

[93] Jahan Ramazani, *A Transnational Poetics* (Chicago, IL: University of Chicago Press, 2009), 67.

[94] Gayatri Spivak, "The Imperative to Re-imagine the Planet" in *An Aesthetic Education in the Era of Globalization* (Cambridge, MA: Harvard University Press, 2012), 335–50, here 340.

3
"The Moon's Corpse Rising"

The Poetic Moon and Imperialist Nostalgia from the U.S. to Kashmir

Beset by England's midlife crisis, Prince Philip nearly crashes his royal plane as he soars into thin atmosphere to get closer to the moon, where the Americans had just landed as self-proclaimed "masters of the universe."[1] "Moondust" (2019), the third-season episode of Peter Morgan's *The Crown* written for the fiftieth anniversary of the Apollo 11 landing, explores British postwar melancholia through the nostalgic symbol of the lost poetic moon.[2] Alarmed by his congregation's search for meaning in "televisions" that broadcast the lunar landing "to 500 million people," the Dean of Windsor tries to save England's soul with poetry, meanwhile imbuing *The Crown* (similarly a television show with millions of viewers) with intellectual seriousness. In his first sermon, the Dean exalts and anglicizes the moon landing by quoting from T. S. Eliot's "Little Gidding":

> We shall not cease from exploration,
> and the end of all our exploring
> will be to arrive where we first started
> and know the place for the first time.[3]

These lines, which are among NASA's favorites to use in its promotional materials, ironically celebrate England, not the America that the poet left: "History is now and England," the poem announces just before the above passage.[4] The

---

[1] According to Loren Eiseley, "Scarcely had the first moon landing been achieved before one U.S. senator boldly announced: 'We are the masters of the universe. We can go anywhere we choose.'" See Eiseley, *The Invisible Pyramid*, (Lincoln: University of Nebraska Press, 1970), 32.

[2] Peter Morgan, writer, "Moondust," *The Crown*, Season 3, Episode 7, directed by Jessica Hobbs, featuring Olivia Colman, Tobias Menzies, and Helena Bonham Carter. Aired 17 Nov. 2019, https://www.netflix.com/watch/80215737?trackId=200257859.

[3] T. S. Eliot, "Little Gidding" from *Four Quartets*, in *T. S. Eliot: The Complete Poems and Plays 1909–1950* (Boston, MA: Houghton Mifflin, 2014), 145. Little Gidding is a key site of the English Civil War, a place also associated with metaphysical poet George Herbert.

[4] For my discussion of NASA's uses of T. S. Eliot in promotional materials, see Chapter 1.

content of the poem is nostalgic, chronicling a return to an origin point; its appearance in this episode also conveys nostalgia for the British Empire. The poem is a relic of a time on the eve of the Second World War when an American citizen expatriated himself to England to immortalize the monarchy in verse—the inverse scenario of Prince Philip's longing to be, three decades later, an unpoetic astronaut. In the Dean's next appearance, he recites lines treasured in the English canon to persuade Prince Philip that the moon landing simply cannot provide the awe and wonder of the Church and poetry. "I'm reminded of Keats," says the Dean: "'What is there in thee, Moon! that thou shouldst move my heart so potently?' Now we know what the moon is: nothing. Just dust. Silence. A monochromatic void." Still, Prince Philip wants to *be* the Apollo 11 astronauts—until he meets them. His searing questioning of the crew of Apollo 11 at Windsor Castle ends not with revelation but rather with Neil Armstrong repeating, "Bang! Bang! Bang! Bang! Bang!" mimicking the sound of the water cooler that kept him awake after his tiring moon walk. The implication is that the Americans have made both the moon and their inherited empire tacky, mechanical, unpoetic. By the end of the episode, Prince Philip has turned his hopes from NASA's moon to the poetry-spouting Dean, who is on his way to infusing the monarchy with meaning again.

"Moondust" exemplifies a widespread phenomenon I identify in this chapter: the expression of national or imperial nostalgia through the dead poetic moon, killed by technoscientific conquest. Nostalgia for the lost moon is an example of what Renato Rosaldo calls "imperialist nostalgia," a "mourning for what [empire] has destroyed."[5] Imperialist nostalgia describes a range of phenomena: it can refer to European yearnings for "traditional" cultures as they were imagined to have been prior to colonization, or to American cultural institutions' sad nods to the noble "vanishing Indian." A closely related concept is Paul Gilroy's "postcolonial melancholia," which describes how "[t]he imperial and colonial past continues to shape political life in the overdeveloped-but-no-longer-imperial countries" in indirect ways, "surfacing only in the service of nostalgia and melancholia."[6] In the 1960s, the moon became a nostalgic object of Anglo-American postcolonial melancholia, a symbolic warehouse of repressed imperial atrocities that continue to shape the present. The Euro-American imagination has long fantasized about lunar domination, sexualizing and feminizing the moon, imagining it to be populated with alien beings evoking racial others, and branding it as the "last

---

[5] Renato Rosaldo, "Imperialist Nostagia," *Representations* 26 (1989): 107–122, here 107.
[6] Paul Gilroy, *Postcolonial Melancholia* (New York: Columbia University Press, 2005), 2.

outpost" in the "final frontier" of space—reviving the doctrine of manifest destiny that drove the mass displacement and genocide of Native Americans in the closing of the frontier. The race to the moon distracted attention and resources from contemporary events registering the legacies of the Anglo-American empire's violent past, including the civil rights movement, the Red Power movement, and escalating conflict in Vietnam. Imperialist nostalgia trained on the moon both expressed and attempted to occlude ongoing manifestations of chattel slavery, land seizures, and overseas conflict at the origin of U.S. imperialism.

Poetry, as *the* genre of the moon, became oddly mixed up in lunar imperialist nostalgia. Poets, too, yearned for the lost poetic moon, in ways that variously sustained and undercut nationalist aims. This chapter demonstrates how lyric poets from the 1960s to the present express nostalgia for the lost poetic moon to rehearse, examine, and sometimes overturn imperialist nostalgia. I begin by tracking lunar imperialist nostalgia in its U.S. contexts. I then consider the work of Kashmiri American poet Agha Shahid Ali, who uses the moon as a figure of national yearning, postcolonial resistance, and transnational mediation. His career documents the Anglo-American imperial processes it was swept up in, as Kashmir's fight for self-determination was influenced first by the British Empire and then by American Cold War geopolitics. Ali's longing for the lost moon of Urdu and anglophone poetry gives way not to a fantasy of national progress but, rather, to elegy for what empire destroys.

3.1 The Lost Poetic Moon

The official poem of the Apollo 11 lunar landing attempts to rekindle wonder for the lifeless, monochromatic moon. On 20 July 1969 the front page of the *New York Times* included Archibald MacLeish's "Voyage to the Moon" beneath an underwhelming headline: "Men Walk on Moon. Astronauts Land on Plain; Collect Rocks, Plant Flag."[7] The headline resonates with the public's sense that NASA's Apollo missions had taken the human species not to a wondrous heavenly body but rather to a dead rock. The Apollo missions had killed off the moon, and MacLeish has trouble resurrecting it:

---

[7] John Nobel Wilford, "Astronauts Land on Plain; Collect Rocks; Plant Flag," *New York Times*, 21 July 1969, https://archive.nytimes.com/www.nytimes.com/library/national/science/nasa/072169sci-nasa.html.

> Presence among us,
> Wanderer in the skies,
>
> Dazzle of silver in our leaves and on our
> Waters silver,
>
> O
>
> silver evasion in our farthest thought—

MacLeish renders the shape of the full moon, alone in a line as though suspended in the night sky, into the O of apostrophic address in a display of hyperbolic metapoesis, restoring the bureaucratic lunar wasteland of the Apollo missions to the enchanted moon of poetry. But the poem swiftly turns from the moon back to Earth in an apostrophic switch that reverses the coordinates of yearning: "O, a meaning! // over us on these silent beaches the bright earth, // presence among us."

MacLeish's poem, like the astronauts in *The Crown*, struggles to summon enthusiasm for the moon. The Apollo 11 astronauts were notoriously unpoetic, to the point that NASA often asked them to try to be more eloquent in press conferences.[8] The crew did have their moments of poetry; Aldrin described the moon's "magnificent desolation" when he landed on its surface nineteen minutes after Armstrong, and Armstrong himself offered a retrospective poem on the tenth anniversary of the lunar landing that assured the public that the moon is not "green cheese" and that it has "a remarkable view / Of the beautiful earth that you know."[9]

Armstrong's and MacLeish's poems are at least as much about the Earth as they are about the moon. "Voyage to the Moon," in fact, is based on *Earthrise* in its depiction of a stunning and vulnerable Earth rising over the lunar wasteland. As I explored in Chapter 1, magazines including the *New York Times* conscripted lyric into presenting *Earthrise* to the public as an emblem of the world-salvific promise of American democracy. The *New York Times*, in fact, had just commissioned MacLeish's essay "Riders on the Earth" for its feature on *Earthrise*, where the essay helped to create planet Earth in an image of white American heteromasculinity.[10] Now contracted to summon awe for the

---

[8] Matthew D. Tribbe discusses the phenomenon of the unpoetic astronaut in *No Requiem for the Space Age: The Apollo Moon Landings and American Culture* (Oxford: Oxford University Press, 2014), 42.

[9] Quoted and discussed in Rivka Galchen, "The Eighth Continent: The new race to the moon, for science, profit, and pride," *New Yorker* (29 Apr. 2019): 46–53, here 53.

[10] Archibald MacLeish, "Riders on the Earth," *New York Times* (25 Dec. 1968), https://timesmachine.nytimes.com/timesmachine/1968/12/25/issue.html.

dead moon of the "Earthrise" era, MacLeish nostalgically returns to Earth. The poem marvels not at the moon but rather at the blue planet, recording how, in killing off the moon, the Apollo missions resurrected the Earth, achingly alive and vulnerable, in an American image.

Critical discussions of *Earthrise* erase the moon, obscuring the object's importance to the first famous shot of Earth from space.[11] In fact, the dead moon helped to invent Earth in this period, introducing a point of scale and contrast against which the planet could rise as a vibrant and cosmically unique entity. And the moon has always been important to the study of Earth. Lunar exploration has a history of decentering and recentering Earth; Galileo's observation of the moons of Jupiter challenged geocentric cosmology, while the race to the moon driven by competing nationalisms led to the establishment of the Planetary Science Institute and, eventually, to the neo-Copernican theory of the Anthropocene.[12] Emphasizing the elided moon of the Earthrise narrative, this chapter scales back from a vision of the whole Earth to ask how and why poets use the moon to reflect on the nation and empire. The moon is curiously absent in the planetary discourses that the Earthrise moment gave rise to. Yet the power of the photograph, and the larger rhetorical creation of "planet Earth" in this particular Cold War moment, depends on the moon—specifically, the moon as a symbol of aspiring nationhood. MacLeish universalizes "American" experience as the human condition in his "Voyage to the Moon." Yet the poem amplifies this nationalist message not by depicting a whole Earth from the vacuum of space but, rather, from the contrasting fixed point of the moon, an emblem of national conquest that is appropriately defeated in the *Earthrise* photograph. MacLeish emphasizes this contrast in the "cold sand" and "silent beaches" of the moon over which the "bright Earth" rises as a "presence among us," the poem concluding on a pronoun evoking a human collective. The plaque left on the moon by Apollo 11 suggests a united species in the vision of peace and harmony MacLeish's poem formally evokes: "We came in peace for all mankind," the plaque reads,

---

[11] Ursula K. Heise points to the tension between ecological narratives and military origins of *Earthrise* and similar photographs in *Sense of Place and Sense of Planet: The Environmental Imagination of the Global* (Oxford: Oxford University Press, 2008), 229. For a postcolonial critique of *Earthrise* and similar photographs, see Tariq Jazeel, "Spatializing Difference beyond Cosmopolitanism: Rethinking Planetary Futures," *Theory, Culture & Society* 28:5 (2011): 75–97. For an ecological treatment of these photographs through the poetry of Seamus Heaney, see Sumita Chakraborty, "Of New Calligraphy: Seamus Heaney, Planetarity, and Lyric's Uncanny Space-Walk," *Cultural Critique* 104 (Summer 2019): 101–34.

[12] For an account of the impact of the space race on the development of planetary science, see Lisa Messeri, *Placing Outer Space: An Earthly Ethnography of Other Worlds* (Durham, NC: Duke University Press, 2016), 5–7.

belying how this imagined global collective came out of the extension of the American military-industrial complex into extraterrestrial space in an attempt to control the world.

Yet *Earthrise*, whatever else it is, is moving, engendering nostalgia in ways that are deeply personal as well as collective. Nostalgia has an endemic geographical component that the distance between Earth and the moon intensifies: as astronaut Frank Borman described the sight of the Earth appearing over the moon, "It was the most beautiful, heart-catching sight of my life, one that sent a torrent of nostalgia, of sheer homesickness, surging through me. It was the only thing in space that had any color to it. Everything else was either black or white, but not the Earth."[13] In her classic study of nostalgia, Svetlana Boym points to the geographical origins of the term in modernity, noting the condition's etymological roots in the Greek for "home" and "longing." Nostalgia originated not in ancient Greece but rather in seventeenth-century Switzerland to describe a "disease" of soldiers, students, and servants working in foreign countries; it is a product of being far away in space as much as in time.[14] Visualizing Earth from the moon, as Borman and MacLeish describe, replaced the moon with Earth as a figure of yearning. Nostalgia, that is, is dialectical, requiring two coordinates: a here and a there, a then and a now, an Earth and a moon.

Earthrise, in fact, is so stirring that it confuses the coordinates of longing: it is at once nostalgic and utopian. Boym describes the similarities between nostalgia and utopia through the suggestive metaphor of a spaceship: "The twentieth century began with a futuristic utopia and ended with nostalgia. Optimistic belief in the future was discarded like an outmoded spaceship sometime in the 1960s. Nostalgia itself has a utopian dimension, only it is no longer directed toward the future."[15] Boym's continuum of longing accounts for how easily nostalgia moved in to fill the cultural-emotional void left by the demise of optimistic humanism so often expressed in astrofuturist works of the 1950s–1960s. Boym argues that space exploration was, ironically, meant to cure rather than induce nostalgia: "More than merely a displaced battlefield of the cold war, the exploration of the cosmos promised a future victory over the temporal and spatial limitations of human existence, putting an end to longing."[16] Yet poetry's continual evocation of this longing—whether it is expressed as utopia or nostalgia—protects the nationalist symbolism of the

---

[13] Quoted in Robert Poole, *Earthrise: How Man First Saw the Earth* (New Haven, CT: Yale University Press, 2010), 2.

[14] Svetlana Boym, *The Future of Nostalgia* (New York: Basic Books, 2001), 3–4.

[15] Boym, *The Future of Nostalgia*, xiv.   [16] Boym, *The Future of Nostalgia*, 346.

moon by repeatedly resurrecting the disenchanted moon as an unrequited figure of yearning so that it remains both possible and interesting to conquer. Because the moon is hard to get to, arriving signifies that a nation has mastered the laws of the universe enough to make the ultimate utopian promise—to cure longing. Nostalgic nationalism couched in the language of futurity drove the space race through John F. Kennedy's rhetoric of manifest destiny in space; by calling the moon the "furthest outpost on the new frontier of science and space" in his "We Choose To Go To The Moon" speech, Kennedy vowed to restore the U.S. to the greatness of its nation-building days.[17] Of course, not all nostalgia is nationalist, and not all nostalgia for a nation is necessarily nostalgic nationalism as galvanized by politicians, who promise the nation's dominant culture that their government will restore the country to an idealized time of their unthreatened hegemony. Lunar poetry in the post-Apollo era has a complex relationship with nostalgia as poets turn to the moon to investigate belonging within and beyond the nation.

Many poets in the 1960s and 1970s captured a sense of encroaching political disenchantment through mourning the demise of the poetic moon. Before Apollo, the moon of Romantic lyric had already been demystified by modernist poets. For T. S. Eliot (writing in 1911), "a washed-out smallpox cracks her face"; for e. e. cummings (in 1923), it is "a fragment of angry candy."[18] But actually going to the moon amplified lunar disenchantment. "Lunicide," in fact, is how poet Edward Lense describes what happened to the moon in the age of Apollo.[19] As Laurence Goldstein put it in his essay on modern poetry and the moon landing on the twenty-fifth anniversary of the occasion, many U.S. poems about the Apollo missions put "an elegiac curse upon the astronauts as the despoilers of a sacred world," dismaying many space program supporters who had expected an enthusiastic response to landing on the most poetic object in the night sky.[20] Babette Deutsch lamented, in the spread of lunar poetry *New York Times* published the day after the moon landing, "Now you have been reached, you are altered beyond belief."[21] Nostalgia for the lost poetic moon was most dramatically expressed by W. H. Auden, who refused

---

[17] John F. Kennedy, Moon Speech, Rice Stadium, Rice University, 12 Sept. 1962, https://er.jsc.nasa.gov/seh/ricetalk.htm.https://er.jsc.nasa.gov/seh/ricetalk.htm.

[18] Eliot, *T. S. Eliot: The Complete Poems and Plays 1909–1950*, 15. E. E. Cummings, *E.E. Cummings: Selected Poems* (New York: Liveright, 2007), 53. For more on the changing place of the moon in modern poetry, see Laurence Goldstein, "'The End of All Our Exploring': The Moon Landing and Modern Poetry," *Michigan Quarterly Review* 18:2 (1979): 192–216.

[19] See Mary Ruefle, "Poetry and the Moon," in *Madness, Rack, and Honey: Collected Lectures* (Seattle, WA, and New York: Wave Books, 2012), 10–31, 16.

[20] Goldstein, "The End of All Our Exploring," 193.

[21] Babette Deutsch, "To the Moon, 1969," *New York Times*, 21 July 1969, 17.

multiple commissions to write a moon poem and wax poetic about the lunar landing. Instead, he wrote a Horatian ode—by hand rather on the typewriter typical of the period—to proclaim, "Unsmudged, thank God, my Moon Still Queens the Heavens."[22] The poem's hyperbolic antiquarianism is a send-up of the imperial techno-utopianism of manifest destiny in space.[23] The Horatian ode traditionally praises an imperial feat; here, Auden inverts its traditional use from praising an empire to praising poetry for its imperial resistance.[24] Even non-poets expressed yearning for the pre-Apollo moon in poetic terms. Journalist Oriana Fallaci, after discovering the "real" moon in a telescope in the 1960s, wrote, "I was very upset. I thought…that I'd never be able to read Sappho or Leopardi without thinking why say it's white when it's black?"[25] Other poets objected to excessive NASA spending that took money away from concerns on Earth. Robert Lowell would, in his sonnet "Moon-Landings," describe the moon goddess Artemis as a "chassis orbiting about the earth," a "void and cold thing in the universe," and link menstruation with the drain of the Apollo program on government spending: "it goes from month to month / bleeding us dry."[26]

Gil Scott-Heron's "Whitey on the Moon" echoed civil rights activists who argued that Apollo spending was especially detrimental to African Americans facing disproportionate levels of poverty.[27] Imagined voyages to the moon have frequently involved white fantasies and fears of racial others. Richard Adams Locke's "Great Moon Hoax" of 1835, for instance, invented winged beings on the moon coded as stereotypically black. In his reading of the "Moon Hoax," poet Kevin Young finds parallels between Locke's racially ambiguous moon dwellers and anti-black, anti-abolitionist, and anti-"amalgamator" riots of the 1830s. Young argues that Locke's hoax presented white

---

[22] W. H. Auden, "Moon Landing," in *Collected Poems*, ed. Edward Mendelson (New York: Random House, 2007), 845.
[23] See Hannah Sullivan, "Still Doing It by Hand: Auden and the Typewriter," in *Auden at Work*, ed. Bonnie Costello and Rachel Galvin (New York: Palgrave MacMillan, 2015), 19–23, here 21.
[24] Auden refused to write the official moon landing poems for the *New York Times*, leading editor A. M. Rosenthal to ask MacLeish. Auden also refused an appearance on the BBC to discuss the landing's philosophical and political importance, asserting that it had none of either. See Edward Mendelson, "'So Huge a Phallic Triumph': Why Apollo Had Little Appeal for Auden," *New York Review of Books*, 12 Aug. 2019, https://www.nybooks.com/daily/2019/08/12/so-huge-a-phallic-triumph-why-apollo-had-little-appeal-for-auden/.
[25] Quoted in Tribbe, *No Requiem for the Space Age*, 189.
[26] Robert Lowell, "Moon-Landings," in *History* (New York: Farrar, Strauss and Giroux, 1973), 185.
[27] Multiple Black poets and civil rights activists in the Apollo era, including Malcolm X and Martin Luther King, Jr., had protested the elaborate spending of the Apollo missions and how they competed with resources of public attention and money to address poverty and environmental concerns that disproportionately affected inner cities. See Neil M. Maher, *Apollo in the Age of Aquarius* (Cambridge, MA: Harvard University Press, 2017), 14.

anxieties about racial others cast as moon dwellers who are "amalgams much like those racial ones feared by 'the common man' opposed to abolition."[28] The real moon landing, Scott-Heron shows, instead planted white beings on the moon transformed into superhuman aliens by their glaringly white spacesuits, paid for by Black people: "Was all that money I made las' year / (for Whitey on the moon?)"[29] Rather than mourning the mythological moon of the Western lyric, Scott-Heron's blues form entangles humor, sorrow, and social critique by merging multiple discourses around the moon, at once a farcical symbol of white American excess and a figure of yearning, a desire for somewhere else and impossible.

3.2 Agha Shahid Ali's Lunar Nostalgia

One of the first poems Kashmiri American poet Agha Shahid Ali published was on the *Earthrise* photograph. Published in Kolkata in 1969 through Lal's Kolkata Writers Workshop, the three-couplet poem combines loco-descriptive conventions with elements of the ghazal, an Urdu form that combines religious and erotic longing with political rebuke:

> Circle these naked rocks, stranger,
> you can't go home again
>
> To these moon-wounds carved in thirsty stones,
> add color with your bleeding feet
>
> Tomorrow, on this hostile horizon,
> your silver earth will rise[30]

Rita Banerjee, one of the few critics to consider Ali as a Cold War poet, calls "Lunarscape" one of his "most dystopian poems" in its presentation of "the recent American moon landing as a neocolonial and self-destructive move."[31] Yet this dystopian poem has an oddly optimistic flair that is part of a

---

[28] Kevin Young, *Bunk: The Rise of Hoaxes, Humbug, Plagiarists, Phonies, Post-Facts, and Fake News* (Minneapolis, MN: Graywolf Press, 2017), 16–17.

[29] Gil Scott-Heron, "Whitey on the Moon," recorded 1970, track 9 on *Small Talk at 125th and Lenox*, Flying Dutchman Records, vinyl LP.

[30] Ali, "Lunarscape," in P. Lal, ed., "Introduction" to *Modern Indian Poetry in English: An Anthology & a Credo*, 2nd edn (Kolkata: Writers Workshop, 1971), i–xliv, here xxxvii.

[31] Rita Banerjee, "Between Postindependence and the Cold War: Agha Shahid Ali's Publications with the Calcutta Writers Workshop," in *Mad Heart Be Brave: Essays on the Poetry of Agha Shahid Ali*, ed. Kazim Ali (Ann Arbor: University of Michigan Press, 2017), 20–32, here 20.

transnational pattern of expressing aspiring nationhood through extraterrestrial imagery. MacLeish's "Voyage to the Moon" and Ali's "Lunarscape" share a preoccupation with nostalgia in the context of imagining national futures. They even end with a neat nostalgic switch of the moon to the now-unattainable Earth as they yearn for the enchanted moon of an old lyric tradition. MacLeish writes the official moon poem into a Western tradition imaginatively extending back to Sappho. Ali's "Lunarscape" yearns for the traditional moon of the ghazal. Both poems, moreover, channel nostalgia for a lost poetic moon into national aspiration. MacLeish's poem summons two millennia of lunar poetry to immortalize an imperial feat: its presence in the *New York Times* invokes the eternal time of poetry and what Jahan Ramazani calls its "long-memoried" forms against the fleeting time of the breaking news story.[32] Ali's "Lunarscape," meanwhile, entered an anglophone Indian literary scene interested in advancing postcolonial India's national aspirations through literary culture that recognized English as a language of India.[33] The poem emerges at the intersection of scientific and literary developments in the 1960s: the rise of the Indian space program and the expansion of a homegrown tradition of anglophone Indian writing.[34] This collision of modernizing forces in the arts and sciences makes it no historical accident that the poet to publish more *Earthrise* poems than any other—a full four volumes of them from 1969 to 1970—was the anglophone Indian poet Satya Dev Jaggi, whom I considered in Chapter 1. Rather than meditate on the shared home of a common humanity, Jaggi and Ali use *Earthrise* to critique the imperialist moon landing and the dislocations brought about by colonization and globalization.

The evocation of vast distances and the skeletal imagery of "Lunarscape" register the aftershocks of the partition of British India in 1947, which led to one of the largest migrations of humans in history and created Kashmir as a disputed territory. In the postwar era, Kashmir represented for people in Ali's affluent and educated class in Srinagar, as well as for much of the world, a multicultural "vale of Paradise." Ali describes the Kashmir of his 1950s childhood as a macrocosm of the culturally expansive ghazal, a place where he grew up "under the *active* influence of three major cultures—Western, Hindu,

---

[32] Jahan Ramazani, *Poetry and Its Others: News, Prayer, Song, and the Dialogue of Genres* (Chicago: University of Chicago Press, 2014), 103, 67.

[33] Lal, *Modern Indian Poetry*, i–xliv.

[34] On the intersection of postcolonial aspirations and the Cold War in the development of the Indian space program, see Asif A. Siddiqi, "Competing Technologies, National(ist) Narratives, and Universal Claims: Towards a Global History of Space Exploration," *Technology and Culture* 51:2 (2010): 425–43.

Muslim."³⁵ In 1989, this fantasy of the harmonious multicultural state was shattered by the outbreak of sectarian violence that has recently escalated; at the time of writing, Kashmir remains the most militarized zone in the world. "Lunarscape" also channels some of the environmental rhetoric around *Earthrise* in recalling violent acts to a specific place. But instead of abstracting environmental devastation to the scale of the whole planet, Ali describes localized violence to the lunar landscape, here figured as a colonial outpost—hardly a stretch given that Kennedy had recently described the moon as the "furthest outpost on the new frontier of science and space."³⁶ The particular colonial outpost described seems to be Kashmir, for which the moon is a national symbol: the flag of Jammu and Kashmir, a princely state under British India from 1846 to 1947, features the star and crescent, marking its predominately Muslim population under Hindu leadership. The silver Earth rising at the end of "Lunarscape" also evokes the optimistic rhetoric of an emerging nation imported from the U.S., a symbol that is oddly out of place in an anti-colonial poem. It will be my concern in the rest of this chapter to account for this poem.

Through much of his career, Ali uses the moon to address the violence in and nostalgia for Kashmir, a place whose cultural productions are rarely understood in Cold War terms. Ali, who identified as a Kashmiri American poet, lived and wrote in the U.S. during and in the wake of the Apollo program. He attended school in Indiana from 1961 to 1963 while his father pursued a Ph.D. in comparative education at Ball State, moved permanently to the U.S. in 1975 at the end of the Apollo program, and became a U.S. citizen just before his death in 2001.³⁷ He frequently plays with the spatial dimensions of nostalgia in his representations of flights, airports, and extraterrestrial space. The moon in particular appears at crucial moments in his work as he negotiates living abroad while Kashmir erupts in sectarian violence; the moon becomes both a symbol of nationhood and a figure of transnational negotiation. The Earth–moon system is, after all, a compelling figure for measuring the distance and relations between two points. As Mary Ruefle suggests, it magnifies the I–thou structure of lyric address: "The moon is the incunabulum of photography, the first photograph, the first stilled moment,

---

[35] Ali, interview by Lawrence Needham, in *The Verse Book of Interviews: 27 Poets on Language, Craft & Culture*, ed. Brian Henry and Andrew Zawacki (Amherst, MA: Verse Press, 2005), 133.
[36] Kennedy, Moon Speech.
[37] Amitav Ghosh, "The Ghat of the Only World: Agha Shahid Ali in Brooklyn," in Ali, *Mad Heart Be Brave*, 199–214, here 207.

the first study in contrasts. Me here—you there."[38] The distance between Earth and the moon evokes the infinite and the intimate, spurring meditations on the distances in both interpersonal and transnational relationships. The trope of longing for home through the moon, of course, predates the Apollo era. One of the poems discovered carved into a wall of the Angel Island Immigration Station by a detained Chinese immigrant in the 1920s is a stanza by Li Bai (701–62) about seeing the moon far away from home and wishing to return:

> Before my bed, the bright moonlight
> I mistake it for frost on the ground
> Raising my head, I stare at the bright moon
> Lowering my head, I think of home[39]

Ali would echo this trope throughout his work, using it to capture the dislocations brought about by accelerating globalization in the postwar era. "The cold moon of Kashmir breaks / into my house," Ali wrote from Arizona as he imagined his family in Srinagar, the coldness of alienation from Kashmir echoed in the choice of "house" over "home." This disconnection is intensified by the line break separating Ali from the family that the moon had earlier risen over almost 8,000 miles away, at once uniting them in its observed westward motion over the Earth and intensifying their separation.[40]

Ali is one of the most candidly and strategically nostalgic of postwar poets. Out of his experiences living between Srinagar, New Delhi, upstate New York, Amherst (Massachusetts), Arizona, and elsewhere, he crafted "a rhetoric of loss, and through loss, the illusion of belonging." As part of this rhetoric, he writes, "I have clung to my loss, to my losses, even to my loss of losses."[41] Unsurprisingly, nostalgia has been a major critical focus of Ali critics, who mostly approach it as a tool of cosmopolitan identification or of anti-colonial nation-building. Bruce King understands Ali's nostalgia within an Indian postcolonial context; he suggests that Ali's poetry exhibits "nostalgia for a lost unified culture" that "has been a feature of Indian Islamic writing from the

---

[38] Ruefle, "Poetry and the Moon," 10–31, here 14.
[39] For a discussion of this inscription, see Jacob Edmond, *Make It the Same: Poetry in the Age of Global Media* (New York: Columbia University Press, 2019), 202.
[40] Ali, *The Veiled Suite* (New York: Penguin Books, 2010), 76. Further references are noted in the text as *VS*.
[41] Ali, "A Darkly Defense of Dead White Males," in *Poet's Work, Poet's Play: Essays on the Practice and the Art*, ed. Daniel Tobin and Pinole Triplett (Ann Arbor: University of Michigan Press, 2008), 144–60, here 148.

mid-nineteenth century, when the British destroyed the society of Delhi after the Mutiny."[42] Shaden M. Tageldin also finds this nostalgia for a "lost unified culture" in Ali, arguing that Ali's nostalgia for Kashmir is a type particular to the postcolonial migrant, describing it as "a longing not for the simple past, but for the past reconstituted and futurized, a past restored to an imagined pre-colonial, pre-exilic integrity and relived, elsewhere."[43] Tageldin describes not quite the "restorative nostalgia" of nationalism but, rather, a reflective consideration of the condition of nostalgia, and how the condition may not just be about holding on to memories but actually inventing experiences and sensations in other times, whether in the past and future or outside of time altogether. King warns against pigeonholing reductive political readings of Ali's nostalgia, arguing that the poetry demonstrates how "[n]ostalgia is not of an ideal world, a place of origins or roots, but of something missed, a past or future, relationships that will not develop, lives he will not have, histories he cannot share except through an extension of the self through desire and imagining."[44] Yet Ali's personal losses almost always reverberate in a collective context; as Boym puts it, "... nostalgia is about the relationship between individual biography and the biography of groups or nations, between personal and collective memory."[45] Ali's "Lunarscape" is an example of this personal and collective collision as the poet apprehends the devastation of partition experienced by his parents' generation. Having been born just after this trauma, he uses nostalgia to connect to memories he was not present for and recasts them futuristically into a space race setting.

For the most part, Ali critics have not dealt with the pernicious sides of nostalgia. One of the most interesting critics of nostalgia in a global anglophone context focuses on the novel but draws conclusions that tap into the nuanced, self-conscious ways in which Ali employs nostalgia while remaining aware of the term's limitations and even dangers. John J. Su argues that many global anglophone novelists "consciously exploit nostalgia's tendency to interweave imagination, longing, and memory in their efforts to envision resolutions to the social dilemmas of fragmentation and displacement," and that "fantasies of lost or imagined homelands do not serve to lament or restore through language a purported premodern purity; rather, they provide a

---

[42] Bruce King, "Agha Shahid Ali's Tricultural Nostalgia," *Journal of South Asian Literature* 29.2 (1994): 1–20, here 6.
[43] Shaden M. Tageldin, "Reversing the Sentence of Impossible Nostalgia: The Poetics of Postcolonial Migration in Salina Boukhedenna and Agha Shahid Ali," *Comparative Literature Studies* 40:2 (2003): 232–64, here 232–3.
[44] King, "Agha Shahid Ali's Tricultural Nostalgia," 3.   [45] Boym, *The Future of Nostalgia*, xvi.

means of establishing ethical ideals that can be shared by diverse groups who have in common only a longing for a past that never was." Salvaging nostalgia from a climate of intense critical suspicion, Su ventures that nostalgia may even "assist ethics" as it "facilitates an exploration of ethical ideals in the face of disappointing circumstances."[46] Ali's later poetry in particular explores how nostalgia might both assist and mark the limits of a qualified cosmopolitan identification based on shared displacement from and longing for a homeland.

In one of his earliest poems written in the U.S., "Postcard from Kashmir" (1979), Ali defines his poetics in terms of nostalgia for Kashmir. This *ars poetica* uses the vast scales of globalization to generate nostalgia, understanding the condition as a confusion of time and space. "Postcard from Kashmir" explores these distortions through the medium of the postcard, a four-by-six-inch object compressed to travel vast distances. The postcard physically shrinks while imaginatively exaggerating the faraway place it comes from, a poetic method of using formal compression to generate emotional expansion. Multiple drafts of the poem show Ali condensing the poem into a compact postcard size, from sprawling draft pages to a version of sixteen lines and then to its final sonnet-like version of fourteen lines,[47] infused with unrequited love for Kashmir:

> Kashmir shrinks into my mailbox,
> my home a neat four by six inches.
>
> I always loved neatness. Now I hold
> the half-inch Himalayas in my hand.
>
> This is home. And this the closest
> I'll ever be to home. When I return,
> the colors won't be so brilliant,
> the Jhelum's waters so clean,
> so ultramarine. My love
> so overexposed.
>
> And my memory will be a little
> out of focus, in it
> a giant negative, black
> and white, still undeveloped.
>
> (*VS*, 29)

---

[46] John S. Su, *Ethics and Nostalgia in the Contemporary Novel* (Cambridge: Cambridge University Press, 2005), 3–4.

[47] Ali, "Postcard from Kashmir," manuscript and typescript drafts (1979). Agha Shahid Ali Papers, Hamilton College Library Special Collections.

With the poet far from home, Kashmir shrinks and blurs, as literal effects of spatial distance and figurative effects of time passing, even as it mentally expands into material for the poet's entire oeuvre. Forecasting nostalgia into the future, he declares that he will draw repeatedly on his memories of Kashmir to develop new poems out of the same photographic negative. "Postcard from Kashmir" depicts the nostalgic object as one that looms at once too small and too large, the highest mountain range in the world shrunk to half an inch even as the sadness of that shrinking becomes an endlessly generative source for poetry.

"Postcard from Kashmir" anticipates Ali's Pulitzer Prize-winning volume *The Country Without a Post Office* (1997), which records the closing of Kashmir's post offices during a period of militancy in the 1990s. The volume mourns a lost vision of Kashmiri cosmopolitanism through astronomical imagery, including the moon. The post office closures inhibited the flows of travel, information, and imagination between Kashmir and the rest of the world depicted in "Postcard from Kashmir." The volume's opening poem likens Kashmir to Osip Mandelstam's "black velvet void," revising the Stalin-era promise "We will meet again, in Petersburg" to "We will meet again, in Srinagar" (*VS*, 171). In addition to Mandelstam, Ali evokes W. B. Yeats, Seamus Heaney, Czesław Miłosz, T. S. Eliot, Ghalib, Faiz, the Koran, the Bible, and other figures and texts of world literature. Stephanie Burt insightfully suggests that Ali turns to global texts and poetic forms in the volume to "counter Kashmir's latter-day isolation."[48] Indeed, Ali's formal poems of the 1990s—including ghazals, villanelles, terza rima, ottava rima, and canzones—explode with both cosmopolitan lineage and cosmic imagery that evoke the etymological origin of the cosmopolitan as a citizen of the universe. Yet Ali's turn to global forms to counter the violently local also critiques empty universalizing rhetoric and mourns the impossibility of the cosmic citizen. *The Country Without a Post Office* is an elegy for a lost ideal of Kashmiri cosmopolitanism rather than a hope that it might be restored in the future. A lunar epigraph taken from the Koran sets up the cosmic-apocalyptic vein of the volume: "The Hour draws nigh and / The moon is rent asunder. (The Koran, Surah 54.1)" (*VS*, 171).

The rent moon evokes the scarred moon of "Lunarscape," where it also signals the inability to return to a home that has been transformed by violence. In the title poem "The Country Without a Post Office," Ali writes in a variation

---

[48] Stephanie Burt, "Agha Shahid Ali, World Literature, and the Representation of Kashmir," in Ali, *Mad Heart Be Brave*, 104–17, here 110.

of ottava rima, which Yeats, to whom Ali refers throughout the volume, used to depict an Ireland gripped by tumultuous political change in "Sailing to Byzantium." Ali's ottava rima variation uses a mirrored rhyme scheme of a, b, c, d, d, c, b, a. The envelope rhyme enfolds the stanza, likening it to a piece of correspondence that cannot be sent:

> Again I've returned to this country
> where a minaret has been entombed.
> Someone soaks the wicks of clay lamps
> in mustard oil, each night climbs its steps
> to read messages scratched on planets.
> His fingerprints cancel blank stamps
> in that archive for letters with doomed
> addresses, each house buried or empty.
>
> (202)

The stanza, encased in the envelope rhyme of "country/empty," evokes the claustrophobia of sectarian violence, creating "that void: Kashmir," a place from which not even news can escape (171). The tiny tip of a minaret points uselessly into cosmic expansiveness; the minaret keeper climbs it in vain each night to read unfathomable astrological "messages scratched on planets." Ali describes the "remains" of the metapoetic minaret keeper's voice as "that map of longings with no limit" (205). Mapping breaks down in cosmic space, where longing continues unbounded without coordinates to anchor it.

The ghazal, the form with which Ali is most associated, frequently deploys cosmic imagery to conjure nostalgia. Burt remarks that "rarely in the history of English has a single poet been so quickly and indissolubly identified with a poetic form."[49] Beyond evincing the tokenism through which an immigrant poet of color gained entrance to the predominately white 1990s New England poetry scene, the ghazal's importance to Ali's career speaks to the enormous political and aesthetic power he granted it as he sought to elegize Kashmir. In one ghazal from the 1990s (poems in the form are usually untitled), Ali meditates on the condition of exile and other types of dislocation by kaleidoscoping examples from Israel, Palestine, Kashmir, Ireland, the Persian tale of the wandering poet Majnoon, and others; the only scale big enough to capture all of these stories of loss is that of extraterrestrial space:

---

[49] Burt, "Agha Shahid Ali," 104.

> In Jerusalem a dead phone's dialed by exiles.
> You learn your strange fate: you were exiled by exiles.
>
> You open the heart to list unborn galaxies.
> Don't shut that folder when Earth is filed by exiles.
>
> ...
>
> Will you, Belovèd Stranger, ever witness Shahid—
> two destinies at last reconciled by exiles?
>
> (*VS*, 297)

This intensely allusive ghazal meditates on the aesthetics and ethics of cosmopolitanism. For Ali, the ghazal was the exemplary form of cosmopolitanism; once a high courtly form, it has, according to Ali, "descendants...not only in Arabic but in...Farsi, German, Hebrew, Hindi, Pashto, Spanish, Turkish, Urdu—and English."[50] This particular ghazal exemplifies how the form challenges the Western "thirst for unity" by valuing *dis*unity (*RD*, 5). Ali explains, "The ghazal is made up of couplets, each autonomous, thematically and emotionally complete in itself: One couplet may be comic, another tragic, another romantic, another religious, another political" (2). As Ali writes, the ghazal's form and thematic range "may bewilder, even irritate, those who swear by neo-Aristotelianism and New Criticism" (6). The ghazal's couplets hold together only through the form's strict and relentless scheme. In a challenge to the temporality of the Western lyric, these couplets could come in any order. The ghazal often ends with a signature of the poet's name, a device that splits and multiplies the figure of voice: the ghazal, often a communally sung form, offers not the presumed unified speaker of the Western lyric but, rather, polyphonic voicing. The contemporary ghazal is not, in other words, closed and symmetrical like the globe but, rather, open, multiple, and fractured, as it weaves multiple, dizzying allusions that cannot be assimilated into a singular shared experience.

The ghazal's sonic rules and scale intensity ignite nostalgia. The soundscape dramatizes the ghazal's thwarted attempts to close in on its object of desire, the Belovèd, who is often God, a lover, Kashmir, or all of these. The ghazal's refrain (*radif*), "by exiles," and the rhyme (*qafia*), "dialed/filed/exiled," occur in both lines of the opening couplet (the *matla*). In ensuing stanzas, they appear only in the second line of each couplet. This pattern throws the ghazal back onto earlier moments in itself that are never quite what they were,

---

[50] Agha Shahid Ali, *Ravishing DisUnities: Real Ghazals in English* (Hanover, NH: Wesleyan University Press, 2000), 12. Further references are noted in the text as *RD*.

offering sonic meditations on nostalgia as an inability to return home. The ghazal also records the scale intensity of nostalgia in its form. Much of the ghazal's effect comes from the tension between irreducible specificity (evoked in the tiniest possible stanzaic form of the couplet) and the infinite, as the ghazal has no maximum number of couplets. As David Ward puts it, "In the bones of the ghazal are two dialectics of the miniature and the vast: one where the brevity of each couplet belies its depth and one between the autonomous couplet and the endless poem."[51] Ali often uses this contrast to imagine nostalgia as a global condition of scalar distortion, as the accelerating speeds and dramatic geographies of globalization can lead the contemporary ghazalist inconceivably far from home. Shadab Zeest Hashmi finds in Ali's scale-intense ghazals a "cosmic nostalgia" as the poet "remembers more than his own exile in his poetry."[52] Cosmic imagery also suits the adorned language of the ghazal and its interest in unattainable divine and erotic love. The moon and wine often travel together in the ghazal through the figure of the Saqi ("one who pours wine," as Ali glosses). In Ali's adaptation of a Ghalib ghazal for his elegiac sequence "From Amherst to Kashmir," he translates a recollection of "those moon-lit nights, / wine on the Saqi's roof— / But Time's shelved them now" (VS, 270). In this nostalgic scene, the moon brings together mysticism and religious desire with erotic, wine-drenched nights. The moon invokes unconquerable erotic yearning through its geographical remoteness, watching over the exploits while its own beauty remains inaccessible.

3.3 The Cold War Ghazal

The moon, according to Ali, is one of the most insistent tropes of the ghazal. In his translator's introduction to the poetry of the socialist Pakistani ghazalist Faiz Ahmad Faiz, Ali writes of the necessity for a translator who doesn't speak Urdu "to learn the nuances of images that would seem too lush to an American poet—images that recur shamelessly in Urdu poetry, among them the moon."[53] In this introduction, he defends the ghazal against exoticization

---

[51] David Ward, "The Space of Poetry: Inhabiting Form in the Ghazal," *University of Toronto Quarterly* 82:1 (2013): 68.

[52] Shadab Zeest Hashmi, "'Who will inherit the last night of the past?' Agha Shahid Ali's Architecture of Nostalgia as Translation," in Ali, *Mad Heart Be Brave*, 183–9, here 183. While Ali liked exile as a "temperamental" descriptor, he rejected it as a political one: "No one kicked me out of anywhere," he often clarified to interviewers eager to play up his biographical strife.

[53] Agha Shahid Ali, introduction to *The Rebel's Silhouette: Selected Poems*, by Faiz Ahmad Faiz (Amherst: University of Massachusetts Press, 1995), xiv.

through the moon. "Lush" and "recur shamelessly" are ghazal-infused descriptions, emphasizing how the form's opulent imagery and intense repetition contribute to its flirtatious tone. Ali's introduction registers that the ornate ghazal does not fit into either the well-mannered, transparent preferences of the New Critical poem or the casual and experimental bent of many free-verse poems. Remnants of British colonial attitudes are at the heart of what presents itself as an aesthetic preference for unadorned poetry in the 1990s U.S. Robert Stilling's recent work traces how the ghazal was nearly eradicated due to homophobic attitudes brought to India by the British that linked homosexuality to "unnaturalness" and "degeneracy." The unattainable Belovèd of the ghazal is always grammatically a male in Urdu and often described in ambiguously gendered terms, making the ghazal a target for homophobic attitudes.[54] Unnaturalness became associated with the ornate artifice of the ghazal form, in contrast to the "natural" poetry of the British Romantics promoted in the colonial schools of British India.[55] (Ironically, the Romantic poets introduced the ghazal form into English, as Ali notes.)[56] Stilling argues that Ali captures this threat to the ghazal in his departures from the form's strict conventions: "[T]he ghazal form has been torn into bits, only to be stitched back together in a new, hybrid form," he postulates. Ali "restor[es] the history of colonial and industrial exploitation to art not by representing it directly but by putting the very brokenness of traditional forms on display, employing the language and motifs of aestheticism to weld back together an endangered verse form while making its near annihilation visible."[57] Not only do Ali's ghazals evoke nostalgia; he often expresses nostalgia for the endangered ghazal form itself—frequently through the moon. In his description of the ghazal moon as "lush" and "shameless" in his introduction to Faiz's work, Ali recalls homophobic threats to the ghazal in British India and suggests that homophobic and xenophobic attitudes in the U.S. underlie the inhospitable response to the ghazal's formal repetition.[58] Ali attempts, in response, to de-exoticize the ghazal, arguing that the moon and other such ornamental symbols are not exotic at all: Faiz has revolutionized them. Ali

[54] For an overview of sexuality and the ghazal, see Mahwash Shoaib, "'The grief of broken flesh': The Dialectic of Death and Desire in Agha Shahid Ali's Lyrics," in Ali, *Mad Heart Be Brave*, 170–82; and Robert Stilling, "Agha Shahid Ali, Oscar Wilde, and the Politics of Form for Form's Sake," in *Beginning at the End: Decadence, Modernism, and Postcolonial Poetry* (Cambridge, MA: Harvard University Press, 2018), 37–87.

[55] Stilling, *Beginning at the End*, 41–2.   [56] Ali, *Ravishing DisUnities*, 83–4.

[57] Stilling, *Beginning at the End*, 83–4.

[58] For a discussion of the relationship between decadence and degeneracy in the history of the ghazal, see Stilling's chapter "Agha Shahid Ali, Oscar Wilde, and the Politics of Form for Form's Sake" in *Beginning at the End*, 37–87.

identifies Faiz as the first to bring "the socialist revolution into the realm of the Belovèd" by suggesting "that waiting for the revolution can be as agonizing and intoxicating as waiting for one's lover."[59] Ali recalls the presence of the socialist revolution in his childhood home in Srinagar, noting that his father often quoted Faiz's elegy for Julius and Ethel Rosenberg.[60] In this elegy, "We Who Were Executed," Faiz politicizes the ghazal moon. The poem portrays Julius longing for Ethel—and uses the "silver light" of the ghazal moon to transform Ethel into an unattainable Belovèd figure who is both politicized and eroticized:

> I longed for your lips, dreamed of their roses:
> I was hanged from the dry branch of the scaffold.
> I wanted to touch your hands, their silver light:
> I was murdered in the half-light of dim lanes.[61]

Here, the Belovèd and the socialist revolution remain equally out of reach, making the desired as inaccessible and alien as the moon.

Most readers understand Ali's interest in Faiz's political ghazals as complex results of British colonization.[62] But Ali's ghazals, and interest in formalism more broadly, should also be understood in terms of Cold War geopolitics. This Cold War context has remained relatively muted in Ali criticism, with most scholars focusing on the aftermath of British imperialism.[63] Ali's poetry, though, is deeply engaged with the Cold War's effects in the postcolonial world. In the introductory note to *The Rooms Are Never Finished* (2001), Ali understands the "large-scale atrocities and the death, by some accounts, of 70,000 people" in Kashmir in the context of worldwide fears of apocalypse; the dispute between the nuclear powers of India and Pakistan means that "Kashmir ... may be the flashpoint of a nuclear war" (*VS*, 245). Understanding Kashmir within Cold War politics helps to make sense of an oddity in Ali's career trajectory, which reverses the expected route of the Cold War anglophone poet. Ali began writing and publishing as a free verse poet in a postcolonial anglophone Indian context that favored formal verse. When Ali moved permanently to the U.S., he turned to formal verse in a climate that

---

[59] Faiz Ahmad Faiz, *Rebel's Silhouette: Selected Poems*, trans. Agha Shahid Ali (Amherst: University of Massachusetts Press, 1995), xx, xiv.
[60] Faiz, *Rebel's Silhouette*, 10.   [61] Faiz, *Rebel's Silhouette*, 39.
[62] Stilling offers the most nuanced of these readings, understanding Ali's attraction to Faiz in terms of postcolonial decadence.
[63] A notable exception is Rita Banerjee, "Between Postindependence and the Cold War," 20–32.

emphasized experimental verse. In both his postcolonial and U.S. contexts, then, Ali's move from free verse to formalism defies expectations. In fact, it reverses the "breakthrough narrative" in U.S. poetry. Initially associated with Lowell and the confessionals, the breakthrough narrative describes a move from tight, restrictive forms into liberating free verse that reflects the reclamation of the authentic self in the therapeutic "breakthrough." Christopher Grobe compellingly argues that this narrative has a Cold War dimension, as poets sought a "violent rupture" with the repressive containment culture that pervaded both U.S. foreign policy and domestic life.[64] Ali's "reverse breakthrough," I venture, also follows a Cold War narrative, as he came to insist on the strictness of the ghazal form to reclaim it from its damaging appropriation into Anglo-American free verse culture.

Ali began as a free verse poet through the Kolkata Writers Workshop led by P. Lal, an advocate of formal verse.[65] Formal poetry was promoted through the New Critical pedagogy that reigned in colonial schools in British India. Moreover, as Nathan Suhr-Sytsma's work on postcolonial publishing implies, because formal lyric poetry carries an enormous amount of cultural authority, writing in the self-consciously "literary" mode of the ideal New Critical poem gave a range of postcolonial and Commonwealth poets entry into a global literary marketplace.[66] Ali follows the trajectory of the postcolonial poet who begins by publishing through a local press and later enters global publication circuits by writing formal lyric poetry; Ali initially published through the Kolkata Writers Workshop, later landing a contract with Norton in New York City with his "formal breakthrough" volume *A Nostalgist's Map of America* (1991). Yet it is noteworthy that Ali did not write formal poetry early in his career when Lal, heading the Kolkata Writers Workshop, supported such verse—particularly as Ali's early work registers a longing to write in forms. Ali's early poem "Introducing" records intersecting imperialisms that drove his early poetic career, from the British Romantic tradition promoted in his Irish Catholic high school in Srinagar, to his British instructors in Kolkata who favored the modernists, to U.S. military operations in South Asia:

---

[64] Christopher Grobe, *The Art of Confession: The Performance of Self from Robert Lowell to Reality TV* (New York: New York University Press, 2017), 27–9.

[65] See Lal, *Modern Indian Poetry*, i–xliv.

[66] Nathan Suhr-Sytsma, *Poetry, Print, and the Making of Postcolonial Literature* (Cambridge: Cambridge University Press, 2017). See especially 23–4 on postcolonial poets and "cultural authority" and 162–95 on the impact of New Criticism-driven Commonwealth universities on Northern Irish poetry.

> At eighteen I was surprised
> by vers-libre. A Ph.D. from Leeds
> mentioned discipline, casually
> brought the waste-land.
> Unawares I was caught in wars
> and wars, Vietnam pulling me
> towards suppleness of language.

Ali concludes the poem, "I lost my chance. / Now I slant my way through rhymes, / stumbling through my twenties."[67] "Introducing" suggests that Cold War geopolitics had discouraged Ali from writing formal verse, with "stumbling" punning on the lack of organizing feet in the poetic line.

In the 1960s and early 1970s, Ali was beginning to write and publish poetry as the CIA was promoting free verse in emerging postcolonial nations in the Global South through organizations like the Congress for Cultural Freedom and the Fairfield Foundation. Through literary journals and festivals in the 1950s–1960s, as Andrew Rubin demonstrates, the U.S. attempted to build a global intellectual network favorable to American values.[68] Suhr-Sytsma, however, takes Rubin to task for over-crediting the expanse and success of this project, pointing out that "[w]riters across the global South—and global North—are always working in, with, and against institutions that permit them only relative autonomy," and these many institutions, local and global, often interrupt one another's processes in ways that are impossible to predict and parse.[69] Studying the impact of CIA initiatives on Mbari publication in 1960s Nigeria, Suhr-Sytsma concludes that affiliated poets' work "was at least partly enabled… by a CIA-funded agency that advanced 'cultural freedom' as preferable to African acceptance of Soviet support." However, he rejects "the assumption that people can do valuable cultural work only when free from determination."[70] Ali, like Suhr-Sytsma's main examples of Wole Soyinka and Christopher Okigbo, was caught up in many competing cultural systems, as he recognized. He ultimately embraced a formalist style at odds with exported American cultural preferences when he emigrated permanently to the U.S. While there is no evidence that the Kolkata Writers Workshop was shaped directly by CIA initiatives, the CIA's promotion of free verse came to

---

[67] Ali, *In Memory of Begum Akhtar* (Kolkata: Writers Workshop, 1979), 13.
[68] See Andrew N. Rubin, *Archives of Authority: Empire, Culture, and the Cold War* (Princeton, NJ: Princeton University Press, 2012), especially 47–73.
[69] Suhr-Sytsma, *Print, Poetry, and the Making of Postcolonial Literature*, 72.
[70] Suhr-Sytsma, *Print, Poetry, and the Making of Postcolonial Literature*, 63.

India in less direct ways through the British educational system. Lal, for one, endorsed the modernist styles of Ezra Pound and T. S. Eliot, on whom Ali would write a doctoral dissertation.[71] As David Caute explains, "the US sponsored a modernist, avant-garde aesthetics that symbolised American political freedoms and that was exported, via CIA-funded magazines, exhibitions and music festivals, to its allies in western Europe"[72]—and then on through British-educated teachers like the "Ph.D. from Leeds."

Just as the U.S. promoted forms "that symbolised American political freedoms," it also exported optimistic symbolism—including the moon. As Leerom Medovoi puts it, "The Cold War imaginary did not divide the globe into two worlds, but rather into three," with the U.S. and USSR both targeting so-called third world or emerging nations, including India and Pakistan. Medovoi traces in Kennedy "an obligatory anti-colonial rhetoric, aimed at newly emergent nations, whose tone of transformative optimism differed sharply from the alarmist containment project of anti-communist discourse."[73] Kennedy described the moon in precisely these terms: "We *choose* to go to the moon," his speech declared to the world, emphasizing how American technological prowess promised freedom and choice. This rhetoric of lunar optimism and national progress grates against Ali's career-long use of the moon as an elegiac emblem. Take, for instance, his elegy for the renowned ghazal singer Begum Akhtar. In it, the ghazal, once a popular form, cries out from its burial in an archive, illuminated by the funereal moon: "Ghazal, that death-sustaining widow, / sobs in dingy archives, hooked to you; / She wears her grief, a moon-soaked white."[74] The lunar poem elegizes Akhtar by *not* using the ghazal, instead eschewing rhyme and embracing heterometric stanzas, recording, as Stilling puts it, the form's "near annihilation" by cultural imperialism.[75]

Ali's late Cold War volume *A Nostalgist's Map of America* (1991) also uses lunar imagery to elegize losses under British and U.S. imperialism. It is set under the vibrant night sky of the Sonoran Desert of southwestern U.S. and northwestern Mexico, long a border zone and a site of cross-cultural encounters. The Sonoran Desert has also been a site of violence against indigenous peoples and ongoing hostility to migrants; Bill Clinton's enactment of the Prevention Through Deterrence policy in the 1990s deliberately created

---

[71] See Lal, *Modern Indian Poetry*, i–xliv.
[72] David Caute, *The Dancer Defects* (Oxford: Oxford University Press, 2003), 3.
[73] Leerom Medovoi, "Cold War American culture as the age of three worlds," *Minnesota Review* 55:7 (2002): 167–86, here 168.
[74] Ali, *In Memory of Begum Akhtar*, 10.   [75] Stilling, *Beginning at the End*, 84.

dangerous border routes through harsh desert landscapes to deter and even kill migrants. The volume criticizes U.S. imperialism as the poet attempts to map the places erased in the making and sustaining of the U.S. empire from the seventeenth to the twentieth centuries. Much of the volume sees the speaker, Shahid, driving through desert landscapes, retracing the bloody routes of Manifest Destiny. Often, he is with his friend and former lover Phil Orlando, whose death from AIDS haunts the volume. In a ghazal style indebted to Faiz's politicization of the form, Ali brings social, political, and erotic losses together by writing near-ghazals, memorializing these losses through a form threatened by the same imperial systems involved in more pernicious acts of cultural erasure. The volume marks Ali's transition from free verse to the tight formalism seen in *The Country Without a Post Office* and later; it frequently references the ghazals of Faiz in both borrowed lines and broken, usually unrhymed, couplets that evoke the ghost of the ghazal. More than in any other volume, the astronomical ghazal imagery gestures to environmental disaster, tapping into the ecological dimension of the Earthrise era. While Ali claimed little interest in "nature," he discussed "the devastation of the earth" as a core political concern of the volume.[76] Nostalgia in the volume often takes on the vast scales of geology and astronomy that eclipse human lifespans and even human apprehension, scales known as "deep time." Ali emphasizes the seas that used to cover the desert, as well as the plants and indigenous cosmologies threatened by climate change. This ecologically inflected nostalgia borders on "solastalgia," a term coined by Glenn Albrecht to describe distress over changing environments, a particular species of nostalgia tied to the Anthropocene.[77] Lunar imagery taps into not only vast geographic scales as in Ali's other work but also the immeasurably vaster scales of deep time to amplify the effects and stakes of nostalgia.

The moon in the volume navigates between scales of loss ranging from a single person to a homeland to the planet; the intensity of grief does not always correspond with how much, quantitatively, is lost. The volume's central elegiac sequence, "In Search of Evanescence," takes its title from Emily Dickinson and, as Mahwash Shoaib points out, positions Ali's work in a queer formalist American tradition that stretches from Dickinson to James

---

[76] Ali, "Agha Shahid Ali: The Lost Interview," interview by Stacey Chase, *The Café Review* 22 (Spring 2011), http://www.thecafereview.com/spring-2011-interview-agha-shahid-ali-the-lost-interview/. As the title indicates, this interview was published long after it took place on 3–4 Mar. 1990.

[77] Glenn Albrecht, "Solastalgia: A New Concept in Health and Identity," *PAN: Philosophy, Activism, Nature* 3 (2005): 41–59, here 41.

Merrill.[78] Ali uses implicitly queer ghazal tropes to attempt to make sense of Phil's loss. "What can I be but a stranger in your house?" Shahid asks Phil's memory in Section 7, evoking the unattainable Belovèd "stranger" of the ghazal (*VS*, 129). Phil's loss tangles with destruction of indigenous cultures. Shahid imagines how the sounds of a dying language "haunt the survivors of Dispersal that country / which has no map" (125). That established map is a palimpsest of lost histories; Phil and Shahid drive over "a highway on the sea / paved by the rising moon," the desert transformed into an earlier epoch under the moon's light; as they pass through tolls, Phil's hand is "rich with planets for coins" (130). The poem weaves the political-erotic knots of nostalgia familiar in the ghazal.

In other poems, the lunar imagery connects distinct localities on the globe. "The moon touched my shoulder / and I longed for a vanished love," Ali writes in "I Dream I Return to Tucson in the Monsoons," a title linking Tucson to South Asia (117). In "I See Chile in My Rearview Mirror," Ali creates a palimpsest of conflicts; as he drives through the U.S., Chile looms large in his mirror as he recalls the 1973 military coup, a Cold War operation to overthrow the socialist President Salvador Allende that resulted in Pinochet's dictatorship. Shahid sees soldiers "firing into the moon," recalling another U.S. Cold War operation: the lunar conquest (161). In "From Another Desert," Ali links the U.S. Southwest to the Middle East through the myth of Laila and Majnoon. The story features the love between Laila and the poet Qays (related to the Arabic word for "moon"), renamed Majnoon ("mad") because his love for Laila drives him insane, sending him wandering the moon-drenched desert longing for Laila (whom he often calls "my moon"). The moon is so important to tales like Laila and Majnoon that the Union of Persian Storytellers wrote a letter to NASA asking that Apollo 11 not land on the moon.[79] Ali's poem also captures the contemporary cultural imperial context by presenting the moon as a figure in pain, and the ghazal as a form in ruins. "From Another Desert" interweaves lines from ghazals by Faiz that Ali was translating while writing the poem. The broken, unrhymed lines demonstrate how the ghazal has been "torn into bits" by imperialism, as Stilling puts it.[80] The poem begins with the "cry of the gazelle when it is cornered in a hunt and

---

[78] Mahwash Shoaib traces Ali's imagination of a queer U.S. poetic lineage in "The Grief of Broken Flesh," 173. On the relationship between James Merrill and Agha Shahid Ali, see Jason Schneiderman, "The Loved One Always Leaves: The Poetic Friendship of James Merrill and Agha Shahid Ali," *American Poetry Review* 43:5 (Sept. 2014): 11–12.

[79] Galchen, "The Eighth Continent," 47. [80] Stilling, *Beginning at the End*, 83.

knows it will die," one of the sources of the ghazal's name (*RD*, 3). The cry is metapoetic, evoking the wounded ghazal in the choppy couplets:

> Cries Majnoon:
>
> Belovèd
>
> you are not here
>
> It is a strange spring
>
> rivers lined with skeletons
>
> (*VS* 139)

This skeletal imagery of the desert also evokes victims of atrocities on the Indian subcontinent, specifically the 1971 Bangladesh genocide committed by the Pakistani military that was exacerbated by U.S.–USSR tensions played out through India and Pakistan. This genocide is the topic of Faiz's "Bangladesh II" that Ali had recently translated, with the line, "The moon erupted with blood, its silver extinguished."[81] Ali echoes that line, fusing frustrated sexual longing with political distress:

> jasmine crushed under
>
> departing feet.
>
> The moon extinguishes
>
> its silver pain
>
> on the window.
>
> (*VS*, 141–2)

A similar lunar moment appears in "The Keeper of the Dead Hotel," which recalls the 1917 Bisbee Deportation in Arizona that kidnapped and removed 1,300 striking miners and their supporters to the desert without water or other supplies. The poem describes

> the silence
>
> of ash-throated men in the desert,
>
> of broken glasses
>
> on the balconies,
>
> the moon splashed everywhere.
>
> (*VS*, 138)

---

[81] Faiz, *Rebel's Silhouette*, 77.

The volume deploys nostalgia to imaginatively recover atrocities of settler imperialism repressed in U.S. national narratives. The volume also responds to emerging violence in Kashmir by placing this violence in a larger transnational context. "Massacres were hushed," as Ali writes in the volume of both the Indian subcontinent and the Sonoran Desert (134).

A Nostalgist's Map of America, of course, demands to be read in terms of nostalgia, which it deploys as a political tool. The title positions Ali's speaker, Shahid, as a nostalg*ist* rather than a nostalg*ic*, evoking a scientific profession as opposed to a sickness and suggesting the professional need for someone positioned as an outsider to the U.S. to map what nationalism must repress to function. In an interview about the volume, Ali describes nostalgia as "a homesickness for what has gone, what has vanished," and adds that "There can sometimes be a homesickness for others' nostalgia for something."[82] "Others' nostalgia" in this volume includes memories and events that have not only faded from personal memory but also been repressed from national memory in the building of the U.S. empire. Ali does this carefully. As Omaar Hena puts it, "Ali frequently returns to the problem of personal and collective mourning in a cosmopolitan frame but does so by putting into tension the difficulties of bridging disjunctive geographies, histories, and poetic forms as they relate to one another contingently and stereoscopically."[83] The poem does not suggest that there is a "universal loss of the past," as Bruce King argues; rather, it explores the loss of a past as experienced specifically by colonized subjects.[84] The poem links these experiences by recording how European imperialism transformed into global capitalism. Passing a town named Calcutta in Ohio, Shahid declares wryly, "India always exists / off the turnpikes / of America" (*VS*, 123). Columbus's misnomer links "Indian Americans" who share little but a common history of imperial domination through a mistake of empire. Ali's encounter with the Tohono O'odham is mediated by the U.S. empire that enables the comparison in the first place. Many of Ali's depictions of indigenous peoples, particularly of the Tohono O'odham, recycle a favorite trope of Rosaldo's "imperialist nostalgia": the noble "vanishing Indian."[85]

---

[82] Ali, "The Lost Interview."
[83] Omaar Hena, "Globalization and Postcolonial Poetry," in *The Cambridge Companion to Postcolonial Poetry*, ed. Jahan Ramazani (Cambridge: Cambridge University Press, 2017), 249–62, here 254.
[84] King, "Agha Shahid Ali's Tricultural Nostalgia," 4.
[85] Rosaldo, "Imperialist Nostalgia," 107.

U.S. settler imperialism has often attempted to "vanish" indigenous subjects through arguments of scientific necessity and, literally, the ability to see the universe more clearly. In 1956, the U.S. built its first national optical observatory, Kitt Peak National Observatory, on sacred land seized from the Tohono O'odham Nation, overwriting Tohono O'odham cosmology with Western astrophysics.[86] In the twenty-first century, protests and activism have clarified the relationship between American astronomy and seizure of indigenous lands: in 2005, the Tohono O'odham Nation filed a lawsuit against the National Science Foundation when it attempted to build the Very Energetic Radiation Imaging Telescope Array System (VERITAS) on Kitt Peak, and the Thirty Meter Telescope protests from 2014 to the present have brought increasing international attention to the U.S. government's ongoing attempts to build telescopes on sacred land on the island of Hawai'i.

Ali likely knew of Kitt Peak's sacred status from his reading of Tohono O'odham history; as he drove through the Arizona-Sonoran Desert, moreover, the gigantic white domes of Kitt Peak National Observatory would have been hard to miss. Whether or not Ali had this contemporary context in mind, his poetry scrutinizes the symbolic role of imperial astronomy in early European contact with the Tohono O'odham. His poem "Desert Landscape" tells the story of Eusebio Francisco Kino, a seventeenth-century missionary, astronomer, and cartographer who lived with the Tohono O'odham and accurately mapped the Sonoran Desert from Mexico into Arizona. The astronomical imagery recalls European imperialism's inaugural move: encroaching on indigenous lands with the goal of epistemologically mastering cosmic space. As Jodi Byrd explains, Captain Cook's first voyage to the South Pacific on the *Endeavour* was conceived of as a scientific mission to chart the transit of Venus in 1769 in the hopes that the event would "unlock the key to the universe's mapping."[87] Byrd's term "imperialist planetarity" describes the alignment of scientific discourse, geographical mastery, and universality that links eighteenth-century European colonization all the way to the extension of

---

[86] In the U.S. government's account, the Tohono O'odham cooperatively leased the land after American astronomers dazzled tribal elders with images of the moon and galaxies seen through a telescope. However, many Tohono O'odham challenged and continue to challenge this story, and recent scholarship has unearthed the financial coercion involved in securing the lease. For the most thorough scholarly account of this coercion, see Leandra Altha Swanner, "Mountains of Controversy: Narrative and the Making of Contested Landscapes in Postwar American Astronomy," doctoral dissertation, Harvard University (2013): 44–91, https://dash.harvard.edu/bitstream/handle/1/11156816/Swanner_gsas.harvard_0084L_10781.pdf?sequence=3.

[87] Jodi A. Byrd, *The Transit of Empire: Indigenous Critiques of Colonialism* (Minneapolis: University of Minnesota Press, 2011), 2.

manifest destiny into extraterrestrial space in the 1960s U.S.[88] "Desert Landscape" imagines the cultural consequences of this missionary work and the Western techniques of mapping indigenous lands, imagining Tohono O'odham women watching their world end; they "can see the moon drown, its dimmed heart gone out / like a hungry child's; can see its corpse rising" (*VS*, 160). The corpse of the rising moon evokes the dead moon of "Lunarscape." Both poems use the ruined moon to depict destroyed homelands, as the volume attempts to position the "vanishing Indian" in broader transnational patterns to mourn shared threats of erasure under empire.

The final poem of *A Nostalgist's Map of America*, "Snow on the Desert," turns from the moon and other astronomical imagery back to Earth. The poem uses cosmic scales to lament violence on the Indian subcontinent, Ali's geographical separation from his loved ones, and the effects of climate change on local desert species and indigenous cosmologies. Ali gives the exact date of the events in the poem, 19 January 1987, a few days after one of the biggest snowfalls in Tucson's history, a rare event signaling effects of anthropogenic climate change. The poet drives slowly over an icy road to take his sister, Sameetah, to the airport for her flight back to Kashmir. They are 7 miles from the airport on a highway where presumably the speed limit is roughly 60 mph, recalling the opening trope of the poem—the idea that it takes light time to travel, so looking into the night sky means looking into the past: "'Each ray of sunshine is seven minutes old'," the poem begins (*VS*, 164). Because they are "seven miles away" from the airport, they should be in sync with celestial motions: seven minutes, assuming a speed limit of 60 mph, from the airport, is the same time it takes the Sun's rays to reach Earth. But climate change throws cosmic harmony out of whack as the unusual snow and fog slow their progress. In this out-of-sync ecosystem, Shahid reflects on an earlier time in geological history: "'Imagine where we are was a sea once. // Just imagine!'" (165). The poem then opens into a reflection on the saguaro cacti, the distinctive 6-foot cacti native to the region that are particularly vulnerable to climate change because of their specific temperature needs. "The desert's plants, / its mineral-hard colors extinguished, / wine frozen in the veins of the cactus," Ali writes, noting that the threat to the saguaros also poses threats to the cosmology and ecology of the Tohono O'odham tribe who use them for material and spiritual purposes. Ali suggests that the saguaros "too are a tribe, // vulnerable to massacre" (164). As Judith Rauscher points out, taking seriously that they are "a tribe...vulnerable to massacre"

---

[88] Byrd, *The Transit of Empire*, xx.

adds an ecological dimension to imperial violence.[89] The poem records how anthropogenic climate change disproportionately impacts the victims rather than beneficiaries of globalization. The poem brings together eco-political consequences of both the British and U.S. empires, expanding the scope of the "violences" in Byrd's remark that it is "time to imagine indigenous decolonization as a process that restores life and allows settler, arrivant, and native to apprehend and grieve together the violences of U.S. empire."[90]

The poem opens into a scale-variable elegy for changing Earth environments, an apocalyptic vision that connects the saguaros, geological history, and Ali's migrant family. The extent of what has changed for the planet's environments becomes a vehicle for apprehending what Shahid has lost in immigrating. The end of the poem constellates ecological, geographical, and personal losses through the ghost of the ghazal form. Ali remembers a night in New Delhi, "perhaps during the Bangladesh War," when Begum Akhtar sang ghazals through a blackout. His memory of her ghazals braids together the volume's disproportionate scales of loss without equating these losses with each other. The memory of her singing in darkness offers a moment in the present

>   to recollect
> every shadow, everything the earth was losing,
> a time to think of everything the earth
> and I had lost, of all
>   that I would lose,
> of all that I was losing.

(VS, 167–8)

The end of the poem locates nostalgia in the past, present, and future. Its omnipresence magnifies the intensity of all that the speaker has lost as he

---

[89] The poem comes out of Ali's reading of a book mentioned in the poem, *The Desert Smells Like Rain* (1982) by Gary Paul Nabhan, an agricultural ecologist, ethnobotanist, and environmental activist with a specific interest in preserving species of plants native to the Southwest and local knowledge of their uses. The book describes the cosmology and sustainable agricultural practices of the Tohono O'odham (the European name, which Ali uses, is "Papago"). As Judith Rauscher writes, "the speaker's perception of the natural world around him is strongly influenced by the Papago view of the desert; or, more accurately, it is influenced by his reading of *The Desert Smells Like Rain*." Rauscher's distinction between "the Papago view of the desert" and Ali's digestion of Nabhan's book on the subject emphasizes Ali's European-mediated encounter with the Tohono O'odham; he also uses the European name for the tribe in the poem. See Rauscher, "On Common Ground: Translocal Attachments and Transethnic Affiliations in Agha Shahid Ali's and Arthur Szu's Poetry of the American Southwest," *European Journal of American Studies* 9:3 (2014).
[90] Byrd, *The Transit of Empire*, 229.

apprehends personal losses through a planetary scale. The broken couplets, which allude to Akhtar's ghazals without embracing the ghazal's rules, evoke these losses at the level of form.

It is through this recognition of nostalgia *for* the ghazal that I wish to return to "Lunarscape." Written twenty years before Ali's turn to formalism and espousal of the "real ghazal," the poem nonetheless contains allusions to ghazal elements that invite reconsideration:

> Circle these naked rocks, stranger,
> you can't go home again
>
> To these moon-wounds carved in thirsty stones,
> add color with your bleeding feet
>
> Tomorrow, on this hostile horizon,
> your silver earth will rise[91]

While Lal praises "Lunarscape" as an example of successful "nontraditional" writing, it has a number of traditional ghazal elements, above all its vast thematic and tonal range.[92] In a signature move of the form, Ali blends sexual desire (evoked in the "naked rocks" and the address to a "stranger," often a stand-in term for the ghazalist's unattainable Belovèd) with religious devotion in multiple traditions (the "bleeding feet" of Christ in his crucifixion and of Hagar in her desert exile). Ali produces a similar effect in a later strict ghazal that protests U.S. post-Cold War militarization in terms that are at once flirtatious, violent, incriminatory, and playful:

> The birthplace of written language is bombed to nothing.
> How neat, dear America, is this game for you?
>
> The angel of history wears all expressions at once.
> What will you do? Look, his wings are aflame for you.
>
> (*VS*, 327)

Mixing Benjamin's angel of history with erotic desire, playfulness ("this game" refers both to bombing and to the game of the ghazal's rules), and serious political rebuke is typical of the contemporary ghazal. "Lunarscape's" flirtatious and politically charged address to the astronaut as a "stranger" is potentially queer, calling out the heteromasculinity of the lunar conquest. Auden

---

[91] Lal, *Modern Indian Poetry*, xxxvii. [92] Lal, *Modern Indian Poetry*, xxxvii.

does the same through the manipulation of poetic form in "Moon Landing": "It's only natural that the boys should whoop it up for / so great a phallic triumph," Auden writes, reserving the longest syllable count of the alcaic stanza to mimic an erection as he questions the desirability of the "natural" by employing a form of high artifice, the Horatian ode.[93] But unlike Auden's formally strict moon poem, "Lunarscape" is missing its signature formal elements. While its couplets are more or less autonomous (Ali does not even bother punctuating between them), this proto-ghazal lacks the rhyme scheme and meter that Ali would later insist on; two decades later in the U.S., he would even write that "a free-verse ghazal is a contradiction in terms" (RD, 2). His omission of formal ghazal conventions in "Lunarscape" intensifies the thematic nostalgia of the poem—the inability to return to the colonizer's home of Earth from the colonized territory of the moon—into formal nostalgia for the lost ghazal and the ravaged homeland it evokes. The moon of "Lunarscape" is elegiac rather than victorious, recording the fate of the ghazal form. In its departures from the ghazal's rules, "Lunarscape" "put[s] the very brokenness of traditional forms on display," as Stilling writes, with the ruined moon mirroring the shattered ghazal form.[94] Ali's free verse ghazal about the moon landing is an elegy for an endangered poetic form and the failed cosmopolitan nation it represents. Gazing from the moon to Earth, "Lunarscape" finds not a utopian future but rather the ghost of a future, as the rising Earth transforms into the corpse of the ghazal moon.

Ali would again visit the Earthrise image in his late-career elegy "From Amherst to Kashmir," which traverses four airports as he travels back to Kashmir to bury his mother's body. The elegy is from his final, formally virtuosic volume *The Rooms Are Never Finished*, its pun on unfinished stanzas (the Italian for "room") expressed in its intricate stanza shapes that are breathlessly enjambed throughout the poem's twelve sections. In Section 3, flying over a Karbala ransacked in the aftermath of the U.S.-led Gulf War, Ali thinks of the violence in Kashmir in fractured stanzas that formally and thematically recall the ghazal. He imagines Begum Akhtar singing "the wound-cry of the gazelle" as his homeland becomes unrecognizable: "PARADISE ON EARTH BECOMES HELL," the newspapers proclaim (257). As the poet nears home on his route from Boston to Frankfurt to Delhi to Srinagar, the stanzas tighten until finally Section 12 closes the elegy outside of the family's house in syllabically strict sestets. The geographic distance traversed by air in the poem

---

[93] Auden, *Collected Poems*, 844.   [94] Stilling, *Beginning at the End*, 83.

opens into lunar distance that signals the unfathomable distance between the living and the dead, who, like the watchful but inaccessible moon, never respond. As relatives congregate to remember his mother in Srinagar, Ali stands at a distance, gazing at the night sky, wishing they would stop discussing "that dark gem of Kashmir"—only to end the elegy for his mother on an Earthrise image that conjures a brutal moment in colonial history:

> ...*How dare the moon*—I want to cry out,
>   Mother–*shine so hauntingly out*
> *here when I've sentenced it to black waves*
> *inside me? Why has it not perished?*
>   *How dare it shine on an earth*
> *from which you have vanished?*
>
> (*VS*, 277–8)

The refracted apostrophe amplifies the distance between Ali and his dead mother. The poet does not simply cry out, metapoetically, to the O of the moon; instead, he addresses the ghost of his mother to say that he *wants* to apostrophize the moon. Even in this most apparently personal moment—an address between the poet and his first home of his mother—the moon is a symbol of the nation. "Black Water" and "black waves" refer to *kalapani*, defined in Ali's endnotes as "the stretch of ocean between mainland India and colonial Britain's most notorious prison on the Andaman Islands. To cross *kalapani* meant being condemned to permanent exile" (379). The recurring ghazal-like rhyme of "out" with "out" emphasizes this state of exile. Like "Lunarscape," "From Amherst to Kashmir" ends with an Earthrise image that evokes the impossibility of a homecoming, channeling the longing of *Earthrise* not into a reflection on the utopian promise of empire but, rather, into an elegy for what it destroys.

4
"Out of This World"
Cosmopolitanism in Cold War Émigré Poetry

On his way home one day in 1948, Czesław Miłosz made a detour to Princeton, New Jersey, to seek the advice of another person displaced by the Second World War: Albert Einstein. Miłosz was grappling with a personal and poetic crisis involving space and time. A visit back to Poland in 1946 had made him confront "astronomical changes" in the postwar nation, and, while working as a diplomatic attaché in Washington, D.C., he struggled with the idea of defecting.[1] But to Einstein the defection question was clear, and his response unequivocal: "You can't break with your country; a poet should stick to his native country."[2] The meeting was a blow. Miłosz had found in Einstein's relativity an antidote to the myopic and dangerous trappings of nationalism. He came to the U.S. as Eastern bloc governments were imprisoning and exiling poets who wrote verse that challenged Stalinist realism, presenting the poet with the choice between being a civil servant writing odes to the state or becoming a martyr for art.[3] Miłosz "worshipped" Einstein as a fellow cosmopolitan émigré and as a thinker who had theorized the kind of poetry he wanted to write, a harmonious "reconciliation between science, religion, and art" beyond what he described as the "prison" of calculable Newtonian time and space that he associated with oppressive systems of thought in both politics and art.[4] Miłosz even adopted Einstein's relativity to describe how exiles and émigrés experience moving through space:

---

[1] Quoted in Adam Kirsch, "Czesław Miłosz's Battle for Truth," *New Yorker*, 22 May 2017, https://www.newyorker.com/magazine/2017/05/29/czeslaw-miloszs-battle-for-truth.

[2] Robert Faggan, *Czesław Miłosz: Conversations*, ed. Peggy Whitman Prenshaw (Jackson: University Press of Mississippi, 2006), 154.

[3] For a discussion of Polish poetry and state strictures, see Clare Cavanagh, *Lyric Poetry and Modern Politics: Russia, Poland, and the West* (New Haven, CT: Yale University Press, 2009), especially 266–76.

[4] Faggan, *Czesław Miłosz: Conversations*, 153–4.

> And space, what is it like? Is it mechanical,
> Newtonian? A frozen prison?
> Or the lofty space of Einstein, the relation
> Between movement and movement?[5]

Einstein's relativity was a cosmopolitan and *poetic* way of thinking—so much so that Miłosz claimed that his cousin, the poet Oscar Miłosz, had independently "discovered" relativity through "intuition," rather than "mathematics," while exiled in Paris in 1916. Einstein's advice to go back to Poland and stay put was just the opposite of what Miłosz imagined relativity promised. He left the Einstein meeting "somewhat numbly" and defected three years later.[6]

Einstein's conviction that Miłosz should return to his "native country" missed an important point: Miłosz did not really have a "native country." Remembered as a Polish poet, Miłosz identified as Lithuanian-Polish; he was born in Šeteniai, a village then under Russian occupation that changed hands across the Polish-Lithuanian border. Miłosz describes this region with otherworldly distance in his poetry and his doomed science fiction experiment, *The Mountains of Parnassus*. One of the characters from that abandoned novel, astronaut Lino Martinez, uses special relativity to imagine the state of exile itself as a home. Martinez muses of his trip to the planet Sardion, "We should have abandoned the word 'journey' the moment we broke through 90 percent of the speed of light.... perhaps the ship resembled a new country, from which there would be no return to the homeland."[7] Miłosz sparks a phenomenon of imagining a cosmopolitan homeland through astronomical discourse that characterizes U.S. Cold War émigré poetry from the 1960s to the early 2000s. Astrophysical propositions, spacecraft, moon landings, and interstellar chill populate the poems of émigrés as analogues for displacement and the longing for a more equitable world. In the absence of a singular, desirable, or even extant national homeland, these poets turn to the lyric as an alternate and preferable world of its own. Indeed, as this chapter demonstrates through the careers of Miłosz and Seamus Heaney, lyric tools create speculative otherworlds uniquely designed to capture the intractable problems of the world we inhabit.

---

[5] Czesław Miłosz, *New and Collected Poems 1931–2001* (New York: Ecco, 2001), 269. Further references are noted in the text as *NCP*.

[6] Faggan, *Czesław Miłosz: Conversations*, 155.

[7] Czesław Miłosz, *The Mountains of Parnassus: A Novel*, trans. Stanley Bill (New Haven, CT: Yale University Press, 2017), 93–4. Further references are noted in the text as *MP*.

4.1 Lyric Otherworlds

In Chapter 1 I noted that Heaney's "Alphabets" (1984) uses the "planetary apostrophe," a phenomenon in which poets invoke the O of Earth seen from space through the "O" of lyric address. I return to "Alphabets" now to explore how Heaney makes the planetary apostrophe into an exploration of lyric cosmology, the origin and design of a lyric poem, to foreground lyric creation as an act of world-making. The poem chronicles his coming of age as a poet, from the preliterate child watching the family name being painted on a gable, to learning the alphabet, to learning his role as a poet, the maker of worlds:

> As from his small window
> The astronaut sees all he has sprung from,
> The risen, aqueous, singular, lucent O
> Like a magnified and buoyant ovum—[8]

What, we might ask, is an astronaut doing in a work about coming of age as an Irish poet? "Alphabets" is a Cold War poem that recycles the trope of the American astronaut as a figure of optimism, the harbinger of a new, resurrected, and united Earth. As with the "silver Earth" rising over the moon at the end of Agha Shahid Ali's "Lunarscape," Heaney's Earth seen from space conveys American optimistic rhetoric of world peace, even as violence in Kashmir and Northern Ireland undercut the notion of a harmonious globe.[9] Unlike Ali, who frequently criticizes American imperialism in his work, Heaney rarely does so, as Justin Quinn notes.[10] But even though Heaney imports a Cold War icon of American salvific hope into "Alphabets," the poem is not as rosy as it might seem. In "Alphabets," rather than using poetry to record the good that exists in the world as it is, Heaney reinvents Earth as a lyric poem.

As a poet frequently in the public eye, Heaney, like Miłosz, was frequently looked to as a "conscience," first in Northern Ireland and later globally. And in his most public appearances, Heaney delivers: "Crediting Poetry," his 1995 Nobel Lecture, teems with optimistic cosmic images; "From the Republic of Conscience," commissioned by Amnesty International, sets the poet up as a

---

[8] Seamus Heaney, *Opened Ground: Selected Poems 1966–1996* (New York: Farrar, Straus and Giroux, 1998), 271. Further references are noted in the text as *OG*.

[9] Ali, "Lunarscape," in P. Lal, ed., *Modern Indian Poetry in English: An Anthology & a Credo*, 2nd ed. (Kolkata: Writers Workshop, 1971), xxxvii.

[10] Justin Quinn, *Between Two Fires: Transnationalism and Cold War Poetry* (Oxford: Oxford University Press, 2015), 143–94.

moral ambassador; and "Alphabets" was itself written for the ceremonial occasion of Harvard's 1984 Phi Beta Kappa address. When performing the public role of poet addressing the polis, Heaney frequently turns to astronomical language to walk a careful line between delivering hope and redressing catastrophe. While always reluctant to speak from the collective perspective of Northern Irish Catholics, he at times summoned the Irish bardic tradition, stretching from the early Irish *filídh* to its twentieth-century revival by W. B. Yeats. The *filídh* were an aristocratic class of poets who were at once magicians, prophets, and lawmakers; Heaney's work, including "Alphabets," often emphasizes poetry's incantatory and spell-like properties.

Miłosz also came from a communal, prophetic, and public poetic tradition; in her work on Heaney and Eastern European poetry, Magdalena Kay describes a shared "deep communal rootedness of the paradigmatic poet in Ireland and Eastern Europe."[11] From their earliest works, Miłosz and Heaney imagine their craft as a vocation with communal power. In his early poem "Campo dei Fiori," Miłosz compares murders of Jews in the Warsaw ghetto to astronomer Giordano Bruno's execution by fire, attributed in folk memory to his belief in the plurality of worlds.[12] The end of the poem evokes poetry's incantatory and magical properties to change reality but casts these as fantastical, in the realm of "legend":

> Those dying here, the lonely
> forgotten by the world,
> our tongue becomes for them
> the language of an ancient planet.
> Until, when all is legend
> and many years have passed,
> on a new Campo dei Fiori,
> rage will kindle at a poet's word.
>
> (*NCP* 34–5)

Heaney's first major poem, "Digging," is similarly an *ars poetica* about the work the lyric ought to do in the world. In it, the poet decides to trade the spade of his ancestors for the pen to write new worlds out of the buried

---

[11] Magdalena Kay, *In Gratitude for All the Gifts: Seamus Heaney and Eastern Europe* (Toronto, ON: University of Toronto Press, 2012), 52.
[12] For more on Bruno's cosmology, heresy, and execution, see Michael J. Crowe, *The Extraterrestrial Life Debate 1750-1900* (Cambridge: Cambridge University Press, 1986). Crowe complicates Bruno's posthumous designation "as a supposed 'martyr for science,'" writing, "It is true that he was burned at the stake in Rome in 1600, but the church authorities guilty of this action were almost certainly more distressed at his denial of Christ's divinity and alleged diabolism than at his cosmological doctrines" (10).

terrain of Ireland's past (*OG*, 4). Heaney, like Miłosz, came to articulate his poetic vocation through astronomy, particularly as he became an increasingly "global" poet near the end of the twentieth century. But what kind of globe do they imagine—and what kind of poetry? Through the cases of Miłosz and Heaney in particular, I will contend that these poets invent lyric otherworlds because there is so much wrong with this one. They task poetry, and their careers, with conjuring a better world through lyric magic, incantation, and spells. Above all, astronomy becomes a way for them to imagine poetry as a vocation and a means of world-making.

The ability to reimagine the world, and in fact to create a fantastic otherworld, is a function long taken to be the domain of narrative. In *What Is a World?*, Pheng Cheah uses the novel to demonstrate how postcolonial and cosmopolitan literature reimagines the spatially conceived capitalist globe as a system of temporal relations that can imagine a more equitable world. "Of all literary forms," he insists, "narrative has the closest affinity to the opening of a world by the gift of time because, as the recounting of events, time is structural to its form."[13] Other cosmopolitan scholarship also features narrative in the remaking of the world. In her readings of transnational fiction, Rachel Trousdale has argued that migrants, exiles, émigrés, and immigrants often invent homelands in narrative in much the same way that I am suggesting Cold War poetic émigrés do. She argues that "[t]ransnational literature is full of fictional countries, alternate histories, and science-fictional worlds because fantastic locations create communities that replace national cultures."[14] Trousdale's focus on how postmodern narration creates fictional countries for exiled writers makes hers implicitly a theory of time, drawing as it does on Benedict Anderson's and Homi Bhabha's classic accounts of how the temporality of narration brings nations and nationalism into being.[15] But the lyric has other, and overlooked, means of capturing the temporality of displacement and instilling new laws of space–time—which is to say, other ways of making a world.

In fact, many of the lyric's strange renderings of time draw from the same tools available to science fiction writers to create new worlds. Seo-Young Chu argues that "science fiction is powered by *lyric*...forces" that are literalized

---

[13] Pheng Cheah, *What Is a World? On Postcolonial Literature as World Literature* (Durham, NC: Duke University Press, 2016), 311.

[14] Rachel Trousdale, *Nabokov, Rushdie, and the Transnational Imagination: Novels of Exile and Alternate Worlds* (New York: Palgrave MacMillan, 2010), 1–2.

[15] Benedict Anderson, *Imagined Communities: Reflections on the Origin and Spread of Nationalism* (London: Verso, 2016); Homi Bhabha, "DissemiNation: Time, narrative, and the margins of the modern nation," in *The Location of Culture* (London: Routledge, 1994), 139–70.

"into ontological features of narrative worlds," such as apostrophe into telepathy or temporal flexibility into time travel.[16] But above all, the modes come together in their manipulation of time. She points out that science fictional and "[l]yric voices speak from beyond ordinary time."[17] Lyric temporality is fantastic in its very conception, its way of imagining time past, present, and future as a bubble of a continuous now; it time-travels and speaks to the dead and the not-yet-living without disrupting its own logic, which is associative rather than narrative. T. S. Eliot opens *Four Quartets* with a description of lyric temporality that deliberately invokes the physicist he wryly called "Einstein the Great": "Time present and time past / Are both perhaps present in time future, / And time future contained in time past."[18] The lyric properties Eliot describes are critical for poets who, coming from other national and cultural traditions, find themselves in the position of aliens in the U.S. in the century in which that country fashions itself as the world. Rather than reproducing the U.S. mythos, these poets create alternate worlds through the tools of lyric temporality. As Miłosz and Heaney show, lyric temporality creates the lyric *itself* as an alternate world with its own laws of physics and reality. This world-making is not usually willful naivety or an escape from material suffering into the transcendent realm of art. Rather, lyric otherworlds critique the world for what it fails to be.

4.2 Cold War Émigrés and the Gravity of History

Imagined as ethical barometers and moral compasses of their home and adopted nations, but unwilling to prettify the atrocities of the past and present, Miłosz and other poetic émigrés frequently turn to the lyric to invent a world that is impossible by the laws of the one we actually inhabit. Miłosz and the self-styled international poets in his legacy—including Joseph Brodsky, Derek Walcott, and above all Heaney—are interested in human collectives and moral absolutes. Yet readers tend to miss in these poets an attendant cosmopolitanism of extraterrestrial despair. The poems do not always offer humanist portraits of an improvable cosmos or human being, nor do they

---

[16] Seo-Young Chu, *Do Metaphors Dream of Literal Sleep?* (Cambridge, MA: Harvard University Press, 2010), 3, 10–11.
[17] Chu, *Do Metaphors Dream of Literal Sleep?* 23.
[18] T. S. Eliot, *T.S. Eliot: The Complete Poems and Plays 1909–1950* (Boston, MA: Houghton Mifflin, 2014), 117. Quoted in Robert Crawford, *Contemporary Poetry and Contemporary Science* (Oxford: Oxford University Press, 2006), 1.

often suggest that poetry can bring about or even model a better society. "No lyric has stopped a tank," as Heaney wrote, and at the height of the Northern Ireland Troubles he saw no way out of the cycle of killings that had afflicted local landscapes for millennia, as he explores in the volume *North* (1975).[19] More often, the poems suggest that cosmopolitan ideals of reason and respect for otherness can prevail only in a world with different natural laws. Part of my aim in this chapter is to reconsider the legacies of Miłosz and Heaney, two influential twentieth-century poets remembered for their steadfast belief in poetry's ability to improve life on Earth. "What is poetry that does not save nations, / or people?" Miłosz asked, and Heaney often cited Miłosz 's example as he sought to write poetry "strong enough to help."[20] Yet lyric poetry, their work insists, is a fantasy. They are read too readily as moralists and have been missed as architects of imagined and impossible worlds.

My claim that Cold War émigré poets see the lyric as a fantastical otherworld might sound just plain wrong. Miłosz, Brodsky, Heaney, and Walcott—male, foreign, aesthetically traditional, Nobel laureates—tend to be lauded as optimists, the consciences of their nations. The literary friendship Brodsky, Heaney, and Walcott forged during their academic appointments in the U.S. enhanced their reputations as the foreign moral compasses for American poetry. All displaced in American institutions in the same generation (Brodsky in exile, Heaney and Walcott by choice), they shared an investment in traditional lyric aesthetics even as they wrote historically grounded poems. They were frequent collaborators who dedicated poems to one another, wrote of one another's political situations, and claimed Miłosz's influence as they articulated cosmopolitan poetic visions beyond the myopia of nationalism. Celebrated in the U.S. for their belief in the sustaining power of art against the depraved conditions of the twentieth century, the trio was embraced by the largely white and privileged U.S. poetry scene of the 1960s and on as history-wizened poets of moral authority, a designation they were at times happy to endorse. Heaney distanced his own communally and historically situated poetry from the solipsistic and ahistorical Anglo-American lyric produced by their "climate of relative freedom and prosperity." This privilege leads, he

---

[19] Seamus Heaney, *The Government of the Tongue: Selected Prose 1978-1987* (New York: Farrar, Straus and Giroux, 1988), 107. The Northern Ireland Troubles were a sectarian conflict in Northern Ireland variously understood as a religious conflict and a hangover from the British colonization of Ireland. The conflict is usually dated from 1968 to 1998, though sporadic violence continues into the present.

[20] Miłosz quoted by Heaney in Dennis O'Driscoll, *Stepping Stones: Interviews with Seamus Heaney* (New York: Farrar, Straus and Giroux, 2010), 290. Further references are noted in the text as SS. Seamus Heaney, *The Redress of Poetry* (New York: Farrar, Straus and Giroux, 1995), 191.

suggests, to "great difficulty in escaping from the arena of the first-person singular" and explains "the irritation we feel at a self-regarding poetry, a poetry of the orphaned self, the enclosed psyche."[21] Heaney made these remarks as anglophone fascination with Eastern European poets like Mandelstam, Herbert, and especially Miłosz swelled in the 1970s–1980s. Walcott, too, wrote of the Eastern European poets with similar amazement. He dedicated his poem "Forest of Europe" (1978) to Joseph Brodsky, where he imagines the two of them in artistic collaboration, "exchanging gutturals in this winter cave." Walcott links them through a shared aesthetic commitment and historical plights that chime across geographies. Walcott connects Brodsky and Osip Mandelstam's imprisonments to both St. Lucia and poetic verse: "The tourist archipelagoes of my South / are prisons too, corruptible"—and yet, "there is no harder prison than writing verse."[22] Quinn challenges this metaphor, noting that these are ethically indefensible lines: the "prison house" of formal poetry, with its strict rules of prosody, is not "harder" than enduring prison. In espousing such sentiments, Quinn writes, Walcott's and Heaney's generation of émigrés "made postcolonialism safe for America." In the 1950s–1960s, Quinn notes, "postcolonialism was considered, and in fact was, inimical to US foreign policy, both soft and hard." But by valuing the aesthetic so deeply, the poetry could "have political content…but by placing such a high value on the act of writing, it supposes a sphere of action irreducible to Cold War conflict."[23] The Brodsky–Heaney–Walcott fellowship styled itself after Miłosz's lines, frequently quoted as a kind of "mantra" by Heaney during the Troubles: "I was stretched between contemplation of a motionless point and the command to participate actively in history" (*SS*, 260).[24] This trio, in the legacy of Miłosz, self-fashioned their careers as a tug between historical duty and aesthetic autonomy, where ultimately ethics and aesthetics could learn to harmonize again.

British and American readers have often understood these poets' careers in these terms. Writing of Eastern European poets' Anglo-American reception, Clare Cavanagh notes that "to writers reared on the Romantic myth of the poet-Christ, the fate of Eastern Europe's modern bards, besieged by history,

---

[21] Quoted in Bernard O'Donoghue, "Heaney and the Public," in *The Cambridge Companion to Seamus Heaney* (Cambridge: Cambridge University Press, 2009), 66–7; Seamus Heaney, "Current Unstated Assumptions," *Critical Inquiry* 7:4 (Summer 1981): 645–51, here 646.

[22] Derek Walcott, "Forest of Europe," in *The Poetry of Derek Walcott, 1948–2013* (New York: Farrar, Straus and Giroux, 2014), 265.

[23] Quinn, *Between Two Fires*, 157, 161.

[24] Heaney quotes from these Miłosz lines in "Away from It All," in *Station Island* (New York: Farrar, Straus and Giroux, 1986), 16.

persecuted by one repressive regime after another, must seem seductive indeed."[25] Indeed, for many Anglo-American poets, "Poland must seem to represent a—somewhat perverse—poetic dream come true."[26] Northern Ireland, Russia, and St. Lucia invoked a similar kind of envy, as Quinn has noted in his study of this crowd in their Cold War contexts.[27] All faced atrocity and emerged with a stronger belief in the power of art, the story goes: wartime brutality for Miłosz, British imperialism and the legacies of slavery for Walcott, exile and hard labor for Brodsky, the bloody sectarian conflict of the Northern Irish Troubles for Heaney. If Miłosz came out of Nazi enemy fire bravely clutching T. S. Eliot's *Collected Poems*, as one of his wartime anecdotes has it, then surely he proved to the West that literature *matters*.[28] Their public personas in the U.S. even involved a missionary flair. Brodsky, a Russian Jewish exile who became a U.S. citizen and poet laureate, encouraged distributing free poetry to the masses for the betterment of American society; one of his ideas was to leave an anthology of modern American verse in hotel rooms like Gideon Bibles.[29] Poetry, under Brodsky's laureateship, would save America's soul.

The culture that received them this way was the same one that aligned Einstein with moral authority in the decades after his discoveries purportedly saved both America and the world. Einstein reoriented scientific and public perception of the entire universe at the same time as he became a beloved American icon, in a postwar moment in which the U.S. identified itself with democracy's universal and world-salvific good. His aphorisms about human curiosity and his wartime plight are at least as culturally familiar as his discoveries about light and gravitation. His photographs are so iconic that they even inspired the face of Steven Spielberg's ultra-empathic character E.T. in the late Cold War film. E.T. is a kind alien being who models for destructive humans what Ursula K. Heise calls a philosophy of eco-cosmopolitanism, in which cosmopolitan sensibility extends beyond regard for other human lives

---

[25] Cavanagh, *Lyric Poetry and Modern Politics*, 2.

[26] Cavanagh, *Lyric Poetry and Modern Politics*, 242.

[27] Quinn, *Between Two Fires*, especially 143–94 for a discussion of Heaney in intersecting postcolonial and Cold War contexts.

[28] Miłosz recounts gripping his copy of T. S. Eliot's *Collected Poems* during a Nazi attack: "Heavy fire broke loose at our every leap, nailing us to the potato fields. In spite of this I never let go of my book," both because it was a University Library book that he did not own himself and because "I needed it." See Czesław Miłosz, *Native Realm: A Search for Self-Definition*, trans. Catherine S. Leach (New York: Farrar, Straus and Giroux, 1968), 249.

[29] Miłosz and Heaney often repeated this anecdote in interviews. Cynthia L. Haven, "A Sacred Vision: An Interview with Czesław Miłosz," *The Georgia Review* 57:2 (Summer 2003): 303–14, here 305; *SS*, 380.

and into "'imagined communities' of both human and nonhuman kinds."[30] Einstein biographies like Helen Dukas and Banesh Hoffmann's *Albert Einstein, The Human Side* emphasize his "modesty, humor, compassion, and wisdom," mirroring descriptions in Miłosz's biographies of the poet's "ethical dimensions," "wisdom," and "compassion."[31] U.S. culture, in fact, embraced the work of its exiles Einstein and Miłosz in remarkably similar ways.

Sometimes, Miłosz endorsed this alignment. "Without sight" (1949), the poem Miłosz dedicated to Einstein following their Princeton meeting, complains of the "empty, oppressive stage set" of Americans' lives, morally elevating the poet and his addressee above a nation that more or less agreed with that assessment.[32] This landscape fusing science and ethics is the same one that Miłosz, Heaney, Brodsky, and Walcott entered into and have been interpreted within. In championing the literary against impossible circumstances, they are often upheld as evidence of literature's innate value and integrity, implicitly backed by the scientific integrity of Einstein's theories that describe consistent laws of the universe. Miłosz adapts his aesthetic "motionless point" from Eliot's "still point of the turning world" in *Four Quartets*—a work that continually registers Einstein's relativity. Miłosz transforms that theory into the moral fiber of the poetic universe, recording the poet's dual debts to what he distinguished as aesthetically autonomous "poetry" and the call of "history."[33] But this ethical system can be realized only in a world of different natural laws, one he imagines through and as the lyric.

Cosmopolitan discourse itself often uses the metaphor of defying the laws of how gravity works on Earth. Kwame Anthony Appiah reminds us that *kosmos* originally carried the sense of "cosmos" or "universe," rather than globe or Earth, and therefore a cosmopolitan is a "citizen of the universe."[34] Against what he describes as the "aery nothing" of abstract belonging to "humanity," Appiah argues for a cosmopolitanism anchored in local attachments and in recognizing rather than attempting to assimilate

---

[30] Ursula K. Heise, *Sense of Place and Sense of Planet: The Environmental Imagination of the Global* (Oxford: Oxford University Press, 2008), 61.

[31] Helen Dukas and Banesh Hoffmann, eds., *Albert Einstein, The Human Side: Glimpses from His Archives* (Princeton, NJ: Princeton University Press, 2016); 1. Andrzej Franaszek, *Miłosz: A Biography* (Cambridge, MA: The Belknap Press of Harvard University Press, 2017), 1.

[32] Czesław Miłosz, *The Year of the Hunter*, trans. Madeline G. Levine (New York: Farrar, Straus and Giroux, 1994), 121–2.

[33] For Einstein's influence on *Four Quartets*, see, for instance, Neville McMorris, *The Nature of Science* (Vancouver, BC: Fairleigh Dickinson University Press, 1989), especially 143; and Katherine Ebury, "'In this valley of dying stars': Eliot's Cosmology," *Journal of Modern Literature* 35:3 (Spring 2012): 139–57.

[34] Kwame Anthony Appiah, *Cosmopolitanism: Ethics in a World of Strangers* (New York: Norton, 2010), xiv.

difference.[35] Heaney describes rooted cosmopolitanism in an image of gravity in his poem "The Birthplace." It returns to the home of Thomas Hardy, amateur astronomer, whose invented rural county of Wessex and attached lexicon become touchstones in Heaney's own locally bound yet expansive poetry, "like unstacked iron weights // afloat among galaxies" (*OG*, 210). Heaney combines a sense of place with the empty, anonymous spaces of the intergalactic vacuum that can consume the émigré. For Appiah, the challenge of cosmopolitanism in the contemporary world is precisely to feel "afloat among galaxies" without adopting an empty universalism that overwrites "human difference."[36] Jahan Ramazani evokes a similar version of cosmopolitanism in his *Transnational Poetics*, also through a metaphor of gravity and spaceflight. Considering poems by W. H. Auden (an earlier émigré in the Brodsky–Heaney–Walcott tradition, who also helped Brodsky to defect), Heaney, and Walcott, Ramazani writes of the "literally *cosmo*politan perspective" such poets adopt in their poems that "peer down on Earth from beyond its surface." Ramazani uses this poetic figure to describe a "rooted cosmopolitanism" similar to Appiah's conception, one that captures the multiple attachments of the transnational poet:

> The figure of the astronaut or airborne traveler might suggest an older model of cosmopolitanism, a claim to universality and detachment from bonds… but more useful for understanding this and other such poems may be concepts of a "located and embodied" cosmopolitanism that aesthetically enacts multiple attachments rather than none.[37]

Walcott himself articulates a shared cosmopolitan vocation through a gravitational metaphor that describes the relatively weak force U.S. poetry norms exert on them: "The three of us are outside the American experience. Seamus is Irish, Joseph is Russian, I'm West Indian.... We're on the perimeter of the American literary scene. We can float out here happily not really committed to any kind of particular school or body of enthusiasm or criticism."[38] Walcott's phrase "float out there" invokes gravity to describe the trio as free

---

[35] Kwame Anthony Appiah, *The Ethics of Identity* (Princeton, NJ: Princeton University Press, 2005), 214.
[36] Appiah, *Cosmopolitanism*, xiv.
[37] Jahan Ramazani, *A Transnational Poetics* (Chicago, IL: University of Chicago Press, 2009), 16–17.
[38] William Baer, *Conversations with Derek Walcott* (Jackson: University Press of Mississippi, 1996), 119.

from the constraints of any particular aesthetic school, a cosmopolitan fantasy of a world republic of letters.

At the same time, Walcott's poetry sometimes records the lack of force the so-called third world exerts on the colonial center. Writing from the U.S. of a homeland shaped by British imperialism, Walcott once described the Antillean poet "as a colonial upstart at the end of an empire, / a single, circling, homeless satellite."[39] The small gravitational force the postcolonial or émigré poet exerts on the imperial center is at once enervating and generative. Walcott also draws on astronomical metaphors to create an archipelagic cosmopolitanism rooted in St. Lucia. Rather than imagining a smooth, interconnected globe with equal access to resources, Walcott imagines the world in Caribbean archipelagic terms of uneven exchanges and hard-to-discern connections under the surface of the water.[40] "The Schooner *Flight*" uses astronomical metaphors to suggest an archipelagic model of the self and its relation to others, held together not by a single gravitational center but, rather, by the forces they enact unevenly on one another. The speaker Shabine, who opens the poem by announcing his complex Caribbean identities that make him "either no one / or a nation," ends his poem by imagining that "this earth is one / island in archipelagoes of stars." The line break after "one" replaces Archibald MacLeish's space age assimilative dream of universal brotherhood with multiplicity.[41] Drawing on Kant's model of "island universes" that coexist and enact force on each other from great distances across the cosmos, Walcott undoes the reductive idea of the "one," or the whole, instead revealing experiences of self and relations with others to be fractured and multiple. This is the opposite vision of the "whole Earth" he critiques in another traveling poem, "The Fortunate Traveller." The poem's persona—a professor of Jacobean literature flying first class on a mission to aid hunger victims—stares down at the world below the plane and reduces it to a "zero," its multiplicity voided into something so reductive that apostrophe, an endemic relational gesture of the lyric, breaks down. The planetary "O" becomes not a figure of relation but, rather, an interpretive void, as the speaker's "iris, interlocking with this globe,

---

[39] Paul Breslin notes that Walcott developed a sense of cosmopolitanism through these poetic friendships, giving him "a quickened sense that his own West Indian experience has parallels elsewhere in the world, in Heaney's Ulster and Brodsky's Russia." See Paul Breslin, *Nobody's Nation: Reading Derek Walcott* (Chicago, IL: University of Chicago Press, 2001), 42.

[40] For a discussion of Walcott's Caribbean archipelago, see Sonya Posmentier, *Cultivation and Catastrophe: The Lyric Ecology of Modern Black Literature* (Baltimore, MD: Johns Hopkins University Press, 2017), 122.

[41] See Archibald MacLeish, "Riders on the Earth," *New York Times*, 25 Dec. 1968, https://timesmachine.nytimes.com/timesmachine/1968/12/25/issue.html.

condenses it."[42] Walcott uses astronomical metaphors to articulate the relationship between the one and the many, developing an archipelagic cosmopolitanism that challenges both the center/margin model of empire in "North and South" and the imperialist "one Earth" rhetoric of "The Fortunate Traveller."

The Brodsky–Heaney–Walcott trio also uses gravity metaphors in *On Grief and Reason*, their homage to Robert Frost, himself an amateur astronomer. Brodsky specifically compares Frost to the gravitational effects on a spacecraft: "One may liken him to a spacecraft that, as the downward pull of gravity weakens, finds itself nonetheless in the grip of a different gravitational force: outward. The fuel, though, is still the same: grief and reason. The only thing that conspires against this metaphor of mine is that American spacecraft usually return."[43] As Russian spacecraft, Brodsky implies, do not. Following his imprisonment and hard labor in the USSR, Brodsky became an exile in the U.S., where he described his displacement in cosmic and gravitational terms. He wrote of exile's pain but also its "pain-dulling infinity, for its forgetfulness, detachment, indifference, for its terrifying human and inhuman vistas."[44] He describes exile in inhuman and indeed interstellar terms in his long poem *To Urania* (the muse, he notes, of astronomy). Exile, Brodsky writes, is like "being hurtled / out of the spacecraft" into a placeless void.[45]

Brodsky goes so far as to define the laws of lyric poetry through the laws of gravity—specifically, in how poetry defies gravity:

> In purely technical terms, of course, poetry amounts to arranging words with the greatest specific gravity in the most effective and externally inevitable sequence. Ideally, however, it is language negating its own mass and the laws of gravity; it is language's striving upward—or sideways—to that beginning where the Word was. In any case, it is movement of language into pre- (supra-)genre realms, that is, into the spheres from which it sprang.[46]

Brodsky traces the lyric's origins to the Greek celestial spheres. Indeed, he offers a cosmology of the lyric, a study in the origin, structure, and fate of

---

[42] Walcott, "The Fortunate Traveller," in *The Poetry of Derek Walcott*, 320.

[43] Joseph Brodsky, "On Grief and Reason," in Joseph Brodsky, Seamus Heaney, and Derek Walcott, *Homage to Robert Frost* (New York: Farrar, Straus and Giroux, 1996), 5–56, here 56.

[44] Joseph Brodsky, "The Condition We Call Exile," in *On Grief and Reason: essays* (New York: Farrar, Straus and Giroux, 1995), 22–34, here 33.

[45] Joseph Brodsky, *Collected Poems in English* (New York: Farrar, Straus and Giroux, 2002), 293.

[46] Joseph Brodsky, "A Poet and Prose," in *Less Than One: Selected Essays* (New York: Farrar, Straus and Giroux, 1986), 176–94, here 186.

poetic design. He develops this idea in his essay on *the* creation myth of Western lyric: the lyre-wielding Orpheus's descent into and emergence from the underworld in pursuit of his lost beloved, Eurydice. Brodsky analyzes Rilke's astronomical rendering of Orpheus's all-consuming grief. In Rilke's passage, Orpheus's "single lyre," with its outpouring of grief for Eurydice, becomes a private cosmic center, akin to "the other earth, a sun / and a whole silent heaven full of stars, / a heaven of mourning with disfigured stars."[47] In his intergalactic reading of Rilke's Orpheus, Brodsky argues that Rilke depicts "a universe ... in the process of expansion" through grief. He explains, "At the center of this universe we find a lyre ... increasing its reach, sort of like the traditional depiction of sound waves emitted by an antenna."[48] Brodsky clarifies that "this astronomy ... is very appropriately far from being heliocentric"; rather, it is an "egocentric" astronomy. This description recalls Bakhtin's dismissal of lyric poetry as a Ptolemaic form, monologic, self-absorbed, and outmoded in the polyphonic modern world.[49] Yet Brodsky is unapologetic about the Ptolemaic lyric, finding in it possibilities for expansion far beyond the self that creates its initial music. By inverting the laws of gravity, lyric poetry can propel its readers and writers *away* from a consuming gravitational center that Brodsky identifies with the ego. Orpheus's lyre "of imagination and mourning" is a big bang event, sending its sound waves outward into a "truly astronomical destination."[50] The lyric keeps the egocentric universe from collapsing in on itself, Brodsky goes on, as "the origin of this centrifugal force enabling him to overcome gravitational pull to any center was that of verse itself." In other words, verse expands the self well beyond the central ego that threatens to swallow everything in its purview. Brodsky's insistence on rigid verse patterns is fully on display in his reading. Recalling the Miltonian iambic cosmos of a "starry dance in numbers that compute," Brodsky argues that patterned verse is the very fabric of the universe. "In a rhymed poem with a sustained stanzaic design," he writes, the process of defying gravitational collapse into the central ego "happens earlier." Simple meter won't do; a full prosodic toolkit is needed to push lyric utterance beyond the self: "In an iambic pentameter blank-verse poem, it takes roughly forty or fifty lines. That is, if it occurs at all. It's simply that after covering such a distance, verse gets tired of its rhymelessness and wants to avenge it."[51] Elsewhere, Brodsky also

---

[47] Brodsky, "Ninety Years Later," in *On Grief and Reason*, 376–427, here 384.
[48] Brodsky, "Ninety Years Later," 409.
[49] M. M. Bakhtin, *The Dialogic Imagination: Four Essays* (Austin: University of Texas Press, 2017), 288.
[50] Brodsky, "Ninety Years Later," 409.   [51] Brodsky, "Ninety Years Later," 410.

links patterned verse to cosmic design: "The seemingly most artificial forms for organizing poetic language—terza rima, sestinas, decimals, and so forth—are in fact nothing more than a natural, reiterative, fully detailed elaboration of the echo that followed the original Word."[52] Brodsky insisted on rigid verse as part of a larger cosmology. Heaney registers this, in his elegy for Brodsky, "Audenesque," written in the trochaic tetrameter of Auden's "In Memory of W. B. Yeats" to chart a lineage in meter.[53] Above all, these poets use gravity to think about the genetic laws of the lyric—its very cosmology—and its relationship to cosmopolitanism. In rearranging those laws they imagine otherworlds, and no poets as intensely as Miłosz and Heaney.

4.3 Miłosz's Lyric Otherworlds

From 1967 to 1971 Miłosz created an otherworld in the more conventional genre for doing so: science fiction. While a tenured professor at Berkeley, he worked on a highly descriptive, mostly plotless, science-fiction novel, *The Mountains of Parnassus*, rejected by his publisher and unknown until it surfaced in his papers at Yale in 2004.[54] Its dystopian elements emerge from Miłosz's disdain for totalitarian regimes, American capitalism, and hippie culture—which is to say, every part of the culture that surrounded him both in Poland and at Berkeley. The novel features a totalitarian planetary state backed by the Astronauts' Union, as a small religious sect organizes a resistance in the Parnassus Mountains. The Astronauts' Union evokes Miłosz's suspicion of patriotic astronauts in the 1960s U.S.: "We were representatives of the species on earth and in the galaxy, and we had been imbued with patriotism and a sense of responsibility for the whole species," astronaut Lino Martinez recalls (*MP*, 88). Beyond that, it's hard to parse a plot. Miłosz's translator, Stanley Bill, anticipates perplexed reactions to the work by underscoring the many perceived divisions between lyric poetry and science fiction. He remarks that picturing the "Nobel Prize"-winner Miłosz "as a science fiction writer is strange and incongruous.... it is difficult to imagine this archetypal European intellectual immersing himself in the world of space

---

[52] Brodsky, "A Poet and Prose," 186.
[53] Seamus Heaney, "Audenesque," in *Electric Light* (London: Faber & Faber, 2001), 64.
[54] Polish intellectual Krytyka Polityczna found the typescript during a fellowship at Yale. For the archival and publication history, see Stanley Bill, "Translator's Introduction: Science Fiction as Scripture," in *The Mountains of Parnassus*, viii–ix.

travel and alien planets."[55] This is not actually so difficult to imagine. Poetry, he declared in the 1980s, "is a kind of exploration of our place in the universe, and precisely not putting in practice any system—even theological systems or scientific systems—but just groping."[56] In his final volume, *Second Space*, he writes, "To be human is to be completely alien amid the galaxies."[57] Indeed, Miłosz's poetry brims with interplanetary references from the 1930s until his death. Why, then, turn to science fiction to address topics his poetry had long taken up?

Rather than turning to science fiction because he found something lacking in the genre of the lyric, Miłosz enmeshed his poetic practice with the genre of science fiction to practice and intensify qualities he was already identifying in the lyric, particularly to do with time and perspective. The very title of the novel suggests generic fusion. The Parnassus Mountains, "so named by nineteenth-century travelers, people of a rather poetic disposition," take their name from the Greek mountain associated with poetry. Despite his publisher's rejection of *The Mountains of Parnassus* and his own misgivings about the success of the novel, Miłosz wistfully suggests, "Who knows whether I might not even have stumbled upon an experimental genre in spite of myself?" (*MP*, 11) Another poem from 1968, "Ars Poetica?," expresses doubt about established genres, opening as an anti-*ars poetica*:

> I have always aspired to a more spacious form
> that would be free from the claims of poetry or prose
> and would let us understand each other without exposing
> the author or reader to sublime agonies.
>
> (*NCP*, 240)

Increasingly, he would discover in lyric temporality a way to make poetry into this "spacious form." Above all, his science fiction novel let Miłosz extend his investigations of how time operates in different genres. Miłosz's lyric poems accomplish many of the same temporal rearrangements and cognitively estranging functions that he experiments with in his science fiction work. For Stephanie Burt, as for most readers, his lyric poems are more successful. Writing of *Parnassus*, Burt notes, "Space travel becomes, at best, a travesty of

---

[55] Bill, "Translator's Introduction," vii. [56] Faggan, *Czesław Miłosz: Conversations*, 176.
[57] Czesław Miłosz, *Second Space*, trans. Robert Hass (New York: HarperCollins, 2005), 58.

the unearthly travel made possible by religious worship and by lyric poems."[58] Miłosz's lyric poems themselves create unearthly worlds.

He does this most insistently in *From the Rising of the Sun* (1973–4). Drafted just after he abandoned *Parnassus*, this long poem seems to be the realization of the "experimental genre" he hoped *Parnassus* had stumbled onto. It is at once a cosmological epic about a homeland and a lyrical collage of quotations from philosophy, folk songs, and sixteenth-century legal documents of the Polish-Lithuanian Commonwealth. Certainly, there is nothing like it. Miłosz conceived of the poem as an experiment with time travel wherein he is not a visitor from the future but an alien visitor from the past. He reflected on the poem: "I speak of myself as a voyager, somebody who comes from another solar system, because time relegated into the past all the cities, all the places I knew in my childhood, my youth. So this is an effort at distancing myself, looking from outside."[59] In *Parnassus*, the astronaut character Karel similarly imagines himself as an otherworldly visitor, but from the future rather than the past. Karel wonders, "Could he cease to be an unearthly spirit observing it from beyond?" (*MP*, 44). Rather than casting otherworlds into the future, as science fiction typically does, *From the Rising of the Sun* casts them into the past, imbuing the past with the speculative qualities more familiar in futuristic vistas. By making the past fantastical, Miłosz makes it newly strange and relevant, worth trying to make sense of rather than relegating to obsolescence.

Section III, "Lauda," is an etymological history of Lithuania that teems with references to astronomy and to poetry. It begins with an epigraph that describes the natural laws of Lauda in ways that are at once precise and deliberately obscure: "A certain eminent alchemist wrote of that country that it is to be found wherever it has been placed in the first and most important need of the human mind." The alchemist suggests that country be named "Parnassus," a nation of poetry (*NCP*, 291). It is both the country of poetry and of Lithuania, a no place or no longer place or never completed place. In the poem's final section, "Bells in Winter," the speaker describes, "under a huge starry sky," the grammar of lost nations accessible only as "an exercise in style": "The pluperfect tense / Of countries imperfective" (326–7). Miłosz describes an impossibility: countries that exist in the present (in Slavic verbs, the imperfective expresses action without reference to its completion) even as

---

[58] Stephanie Burt, "Space Travels," *The Nation*, 1 June 2017, https://www.thenation.com/article/archive/Czeslaw-Miloszs-space-travels/.

[59] Cynthia L. Haven, *Czesław Miłosz: Conversations* (Jackson: University Press of Mississippi, 2006), 60–1.

they have already ended (the pluperfect). The grammar is that of exile when the exile has left a place that continues to exist only in memory. By addressing this situation in lyric rather than narrative time, Miłosz delves into the impossible temporality of exile. Using the flexible and otherworldly temporality of lyric poetry, he visits the past and infuses it with the strangeness science fiction typically projects onto the future.

Anglo-American readers have largely missed this strangeness in Miłosz. By way of example, we might take "Incantation" (1968), often read in weighty U.S. civic contexts, including the wake of 9/11. Miłosz wrote "Incantation" at the height of the space race and as he was working on *The Mountains of Parnassus*. Fantastical elements permeate the work, from strange temporalities to a unicorn:

> Human reason is beautiful and invincible.
> No bars, no barbed wire, no pulping of books,
> No sentence of banishment can prevail against it.
> It establishes the universal ideas in language,
> And guides our hand so we write Truth and Justice
> With capital letters, lie and oppression with small.
> It puts what should be above things as they are,
> Is an enemy of despair and a friend of hope.
> ...
> Beautiful and very young are Philo-Sophia
> And poetry, her ally in the service of the good.
> As late as yesterday Nature celebrated their birth,
> The news was brought to the mountains by a unicorn and an echo.
> Their friendship will be glorious, their time has no limit.
> Their enemies have delivered themselves to destruction.
>
> (*NCP*, 239)

The poem seems straightforward enough: human reason, epitomized in the union of philosophy and poetry, will prevail over the horrors of history. "Incantation" epitomizes why Miłosz is upheld as a figure of conscience and a successful apologist for poetry. However, not all is as it seems. "Incantation" experiments with time to infuse the poem with fantastical elements that deliberately bring its believability into question. Miłosz, in fact, was specifically thinking about the importance of time to fantastical world-building in this period. In his translator's introduction to *Parnassus*, Bill includes excerpts from an obscure Miłosz essay, "Science Fiction and the Coming of the

Anti-christ," in which the poet argues that both prophecy and science fiction "use the same basic grammatical conceit to make their prognostications of the future more plausible." Bill quotes Miłosz:

> We would include here any narrative that pretends to be written in the past tense, whereas it should have been written in the *futur accompli*; it should have been, but cannot be, because grammar itself stands in the way.... A prediction...is disguised grammatically: a hero living in the year 3000 "did" and "went." But we find the same thing in the Revelation of St. John: that which is predicted is told as something that has already occurred—in a vision on the island of Patmos.[60]

Miłosz invents a grammatical tense that does not exist to explain the temporality of science fiction and religious prophecy. He plays with the *futur antérieur*—an action that will have been completed in the future, akin to "They will have left [by then]." The *futur accompli* would be an action already accomplished in the future, a tense that doesn't exist, as in, "They left [3,000 years in the future]." Where science fiction and prophecy manipulate this tense to make their worlds more believable, Miłosz manipulates the tenses of "Incantation" to make the world it describes *less* believable, much harder to place. "The news" of the alliance of poetry and philosophy "was brought" in a vague time, "[a]s late as yesterday" rather than "as recently as yesterday." The speaker, it seems, is thinking of "yesterday" from the perspective of the past rather than from the perspective the present. The news "was brought" by a supernatural creature and an echo, itself a temporal term that means the original sound has already passed. The declaration that "their time has no limit" is undercut by the future tense of their friendship, which after all is limited to the future that the poem does not show us. And *when* have "their enemies...delivered themselves to destruction"? The tenses are contradictory, specific, and vague all at once. It is impossible, in ordinary time and space, to locate this world governed by reason and affirmed in the union of poetry and philosophy.

4.4 Heaney's Lyric Otherworlds

Heaney, reading this poem at Berkeley in 1970 as the Troubles erupted in Northern Ireland, was compelled by this argument for poetry's importance in

---

[60] Bill, "Translator's Introduction," xvi–xvii.

times of duress—but not because he thought the poem offered a path to a better reality. He quotes "Incantation" in full at the beginning of his influential late Cold War essay "The Impact of Translation" (1988), where he declares that "poets in English have felt compelled to turn their gaze East and have been encouraged to concede that the locus of greatness is shifting away from their language."[61] Heaney's enthusiasm for and promotion of Miłosz and other Eastern European poets in this era was inflected by both his Irish and his U.S. contexts. He notes that Eastern European poets' situations make them "attractive to a reader whose formative experience has been largely Irish," due both to experiences of political pressure on their art and their communal orientation.[62] Yet Heaney did not read Miłosz in earnest until he began to hold academic appointments frequently in the U.S., where he experienced similar dismay at the individualistic American lyric scene. Alienated at various times by experimental verse, confessional verse, and New Critical decorum, Heaney turned to Miłosz to renovate his conception of lyric poetry, as Kay has convincingly shown.[63] But his first encounter with the work brims with specifically Anglo-American enthusiasms for samizdat poetry from behind the Iron Curtain. Heaney recounts the attraction "Incantation" held for him when Robert Pinsky read it to him in the upstairs privacy of his home in Berkeley. Slyly borrowing language of censorship and resistance, Heaney recounts the "altogether thrilling" and "conspiratorial" experience, even a "feeling of collusion," in "enjoying a poem that did things forbidden" in modern Anglo-American ideas of the lyric. Heaney praises the "unabashed abstract nouns and conceptually aerated adjectives" that violated Pound's "go in fear of abstraction." Even the lyric voice does the unexpected, and this poet was clearly delivering "a message" disallowed in poetry: "Now here they were in a modern poem—big, pulpit-worthy affirmations, boosted all the further by that one metaphorical flight about a unicorn and an echo in the mountains, and fortified by the hovering irony of the final line." Heaney is one of Miłosz's only English-speaking readers to note the irony of that final line—"Their enemies have delivered themselves to destruction"—which after all grates against the poem's promise of universal good. Heaney noted something else about the poem that has escaped most of its commentators: the poem does not pretend that any of its moral affirmations are real or achievable. He points to the title:

---

[61] Seamus Heaney, *The Government of the Tongue: Selected Prose 1978-1987* (New York: Farrar, Straus and Giroux, 1989), 38.
[62] Heaney, *The Government of the Tongue*, xx.   [63] Kay, *In Gratitude for all the Gifts*.

This is a spell, uttered to bring about a desirable state of affairs, rather than a declaration that such a state of affairs truly exists—for nobody knows better than the author how long and how invincibly the enemies of human reason can prevail. What gives the poem its ultimate force is, therefore, the intense loss we recognize behind its proclamation of trust.[64]

"Incantation" does not suggest that poetry can bring about a world governed by human reason, nor even exactly that poetry records or affirms the good that persists amid evil in the world. Instead, the poem records an elegiac desire to bring a better world into existence through incantatory magic— elegiac because the poem relegates this possibility to a hazy past. Stephanie Burt, attuned like Heaney to Miłosz's speculative poetics, argues that "Incantation" deliberately describes a world that is impossible to believe in: "only in a certain willful, subjunctive sense can the poem believe it: its antique machinery of allegory—not reason but talking reason, personified, and eventually Philosophia—says what modernism in the arts, and modern history with its gulags and its famines, suggests that we can no longer believe."[65] Miłosz's poetry is distinctive in how it "allows us to envision another world, one that can help us because it is not our home; one where justice prevails alongside unicorns."[66] In other words, Miłosz suggests that justice is a fantasy that lyric poetry poignantly records our inability to believe in.

Miłosz's poems offer utopian fantasies as means of world-making. China Miéville defends the fantasy mode in "Cognition as Ideology," arguing that fantasy offers a powerful critique of the world through estrangement—an effect that Darko Suvin specifically identifies as the work of science fiction. Exploding Suvin's genre containers, Miéville uses the fantasy mode to show how we might cluster many genres under the rubric of "the literature of alterity."[67] His work opens the possibility that poems, too, do the work of fantasy when they present a world that is radically different and difficult to comprehend, offering an encounter with radical alterity rather than a description of how the world we inhabit might actually come to improve. In a similar

---

[64] Kay, *In Gratitude for all the Gifts*, 37.
[65] Stephanie Burt, "Incantation," in *The Poem is You: 60 Contemporary American Poems and How to Read Them* (Cambridge, MA: The Belknap Press of Harvard University Press, 2016), 63.
[66] Burt, "Incantation," in *The Poem is You*, 64.
[67] China Miéville, "Cognition as Ideology: A Dialectic of SF Theory," in *Red Planets: Marxism and Science Fiction*, ed. Mark Bould and China Miéville (Middletown, CT: Wesleyan University Press, 2009), 231–48, here 234. Darko Suvin discusses science fiction and cognitive estrangement in *The Metamorphoses of Science Fiction: On the Politics and History of a Literary Genre* (Bern. Switzerland: Peter Lang, 2016), 7–9.

vein, Fredric Jameson argues that "utopia as a form is not the representation of radical alternatives; it is rather simply the imperative to imagine them."[68] Heaney's utopian poetry, like Miłosz's, has elements of the fantastic. One well-known example occurs in his version of Sophocles' *The Cure at Troy* (1990), which Bill Clinton imported into a real-world political conflict when he quoted from it in his speech in Derry, Northern Ireland, during the peace process:[69]

> History says, Don't hope
> On this side of the grave,
> But then, once in a lifetime
> The longed-for tidal wave
> Of justice can rise up
> And hope and history rhyme.
>
> (*OG*, 306)

Of course, "hope and history" do *not* in fact "rhyme," not even by the poem's own craft laws. They share only their first letters (and an *o*), recalling Brodsky's declaration to Heaney that "the only thing politics and poetry have in common is the letter p and the letter o."[70] There will never be a world in which they do rhyme; even the natural laws of the poem's prosody ask us simply to imagine that these concepts could harmonize. As in Miłosz's "news" of human reason "brought to the mountains by a unicorn and an echo," the poem goes on to liken the possibility of peace to "miracles," "cures," and "healing wells." The poem then meditates on its own making, the work that went into imagining an alternative world where "justice" prevails, one that is decidedly beyond this Earth: "Here on earth my labours were / The stepping stones to upper air" (*OG*, 307). In his Nobel speech, Heaney likens the stage in Stockholm not to a "stepping stone" but rather to a "space station" where, "for once in my life, I am permitting myself the luxury of walking on air." He goes on, "I credit poetry with making this space-walk possible" (*OG*, 417). Throughout his career, he would attempt to make the laws of the lyric defy those of the gravity of history.

---

[68] Fredric Jameson, *Archaeologies of the Future: The Desire Called Utopia and Other Science Fictions* (London: Verso, 2005), 416.

[69] Bill Clinton, "Remarks to the Community in Derry, Northern Ireland," 30 Nov. 1995, https://www.govinfo.gov/content/pkg/PPP-1995-book2/pdf/PPP-1995-book2-doc-pg1809.pdf.

[70] Joseph Brodsky and Seamus Heaney, "Poetry and Politics: A Conversation between Seamus Heaney and Joseph Brodsky," ed. Fintan O'Toole, *Magill*, 31 Oct. 1985, https://magill.ie/archive/poetry-and-politics-conversation-between-seamus-heany-and-joseph-brodsky.

No single poet has developed astronomy into the tools of a poetic vocation as thoroughly as Heaney, yet thinking of Heaney as an astronomical poet might seem incongruous. His early poetry in particular delves into the rural, material ground of his native Derry, Northern Ireland. He is also associated with Northern Ireland through the Belfast Group, a cluster of Northern Irish poets including Michael Longley, Derek Mahon, and Paul Muldoon, who rose to prominence during the Troubles as part of the so-called Northern Irish Renaissance. Longley, one of Heaney's closest collaborators, understands Heaney's fascination with astronomy to be career-defining. In his satirical poem "The Group," which presents members of the Belfast Group as comically obscure Greek poets, Longley describes Heaney as Ion of Chios, "Who specialises in astronomical phenomena."[71] Indeed, beginning in 1970 as he spent substantial time abroad, Heaney's poetry and essays teem with *Star Trek* references, moon landings, stars, galaxies, spacewalks, and spacecraft. In "The Milk Factory" (1987), Heaney compares the factory floor to "a bright-decked star ship" (*OG*, 291). He also compares farm labor to interstellar astonishment in "The Pitchfork," which elegizes his father through the image of a pitchfork in outer space, "Its prongs starlit and absolutely soundless" (320). In one interview, Heaney compared reading lyric poetry to the "buoyancy" in a "spacecraft," and noted that he sought "to make the language do something akin to beaming it up, like in *Star Trek*."[72] Like Brodsky, Heaney imagines the lyric as defying the laws of gravity. It offers an otherwise impossible experience of buoyancy, levity, and, occasionally, outright levitation, crucial experiences during intractable atrocities. Heaney's early work captures the despair of the Northern Irish situation through the metaphor of gravity. In "Kinship," he writes of coming of age in the North:

> I grew out of all this
> like a weeping willow
> inclined to
> the appetites of gravity.
>
> (*OG*, 118)

In his Nobel speech, Heaney reflected on the difficulty of becoming a poet "out of all this" in Northern Ireland, where he had "to adjudicate among promptings variously ethical, aesthetically, moral, political, metrical, sceptical, cultural,

---

[71] Michael Longley, "The Group," in *Collected Poems* (London: Jonathan Cape, 2006), 321.
[72] Seamus Heaney, "A soul on the washing line," *The Economist*, 5 Sept. 2013, https://www.economist.com/prospero/2013/09/05/a-soul-on-the-washing-line.

topical, typical, post-colonial and, taken all together, simply impossible" (*OG*, 418). Largely out of his relationships with Miłosz and Brodsky, Heaney develops a career-spanning metapoetic metaphor of gravity, one that maps onto his career trajectory from a poet rooted in the earth to a poet of the "marvellous" recorded in his 1991 poem "Fosterling" (331).

Heaney first started using the gravity metaphor when and as Brodsky did: likening life in the U.S. to a low-gravity experience. Cynthia L. Haven recalls that Brodsky "commented that living under Soviet rule was akin to living under increased gravity, when every word and act had enormous repercussions. Living in the West, in contrast, was like living on the moon—one could bounce and somersault with no effort, but it meant nothing, since everyone else was doing exactly the same."[73] Heaney's very first use of the gravity metaphor, while a visiting professor at Berkeley in 1969–70, similarly compares being in the U.S. to the "loosening gravity" of the moon. Writing from Berkeley the year after the Apollo 11 moon landing, as Miłosz was writing nearby about the Astronauts' Union of *Parnassus*, Heaney also began to imagine leaving Earth's gravitational pull, writing of the lunar "loosening gravity" he felt in California as sectarian violence escalated in Northern Ireland.[74] In its dual longing for and misgivings about the freewheeling 1960s, the poem examines the poet's public and communal responsibilities while insisting on the autonomy of the poetic enterprise.

The poem uses the trope of lunar nostalgia I explored in Chapter 3 to contemplate homeland, history, and empire. Sitting under a map of the moon in his borrowed office a few months after the moon landing, he reflects on the appearance of the moon the night before his flight out from Donegal and registers a duty to turn back to Ireland from California, at least in his poetry, to address the violence he has left behind for the "free fall" of the American West:

> I sit under Rand McNally's
> 'Official Map of the Moon'—
> The colour of frogskin,
> Its enlarged pores held
>
> Open and one called
> 'Pitiscus' at eye level—
>
> (*OG*, 76)

The Rand McNally publishing company (known for publishing terrain maps) was founded in 1856 by William Rand, who partnered with Andrew McNally,

---

[73] Haven, *Joseph Brodsky: Conversations*, ix.   [74] Haven, *Joseph Brodsky: Conversations*, 76.

a famine-era immigrant from County Armagh in Northern Ireland. This provenance links Heaney's own dislocation from a troubled Ireland to the lunar map. Heaney's description of the U.S. map of the moon evokes the British colonial ordnance maps that anglicized local place names in an act of cultural annihilation, severing the relationship between language and landscape. The moon's surface is likened to a foreign territory exposed in transparent detail so that the colonizer can master the landscape both epistemologically and geographically, with the crater's comparison to an "enlarged pore" that sits "at eye level" intensifying the sense of surveillance. The linguistic topography of the map, captured in the crater's name of "Pitiscus" (after the German astronomer), invokes the anti-colonial *dinnseanchas* revival in twentieth-century Irish poetry. "Westering" roots itself in the Irish landscape through an old Irish-language form employed by the *filidh*, the *dinnseanchas* tradition, the recitation of local place names and events associated with them in verse. Heaney often plays with Irish-language words and their physical relationships to the landscape in this tradition, as in his place-name poem "Broagh," where the O of a conquering heel print made in the landscape graphically recalls the vowel of the place name (*OG*, 55). In reciting and recuperating these place names, Heaney summons the supernatural dimension of the *filidh* tradition to bring into being a world that has vanished. Moving the *dinnseanchas* tradition onto the moon, an otherworld entirely, Heaney portrays the U.S. lunar conquest as an imperial event. With the open "frogskin" of the map, Heaney depicts imperial cartography as an autopsy, an act that exposes and dissects the body of a foreign creature or place for the benefit of the rational, enlightened conqueror. The frogskin also recalls the threatening frogs of Heaney's early poem "Death of a Naturalist," which are "cocked" and "poised like mud grenades" (*OG*, 5). The lunar map implicitly links British imperialism to American imperialism through their shared interest in cartographic violence.

Heaney wrote "Westering" while weighing a move from Belfast to the Republic of Ireland to "[escape] from the massacre," a widely criticized decision that some Northern Irish critics took as a betrayal.[75] Heaney returned from Berkeley in August 1971, the same week the British Army enacted internment in Northern Ireland. The poem uses the moon landing and the freewheeling American ideology it represented to work through ethical commitments to homeland and history. Its title change from "Easy Rider" (after Dennis Hopper and Peter Fonda's 1969 hippie film about two drug-dealing

---

[75] Heaney uses this phrase in "Exposure," *OG*, 136. For the most searing criticisms of Heaney's treatment of the Northern Irish political situation, see Ciaran Carson, "Escaped from the Massacre?", *The Honest Ulsterman* 50 (Winter 1975), 183–6; and Edna Longley, " 'Inner Emigré' or 'Artful Voyeur'? Seamus Heaney's *North*," in *Poetry in the Wars* (Newcastle upon Tyne, UK: Bloodaxe Books, 1986).

bikers) to "Westering" (after its more solemn intertext, John Donne's "Good Friday, 1613, Riding Westward") registers the pull between the individual autonomy promised by American culture and the duty of collective responsibility.[76] "Westering" concludes in devotional language that recalls both John Donne's travel away from religious duty in "Good Friday, 1613, Riding Westward" and the poet's duty to address the violent subjugation of the Catholic population in Northern Ireland:

> Under the moon's stigmata
>
> Six thousand miles away,
> I imagine untroubled dust,
> A loosening gravity,
> Christ weighing by his hands.

(OG, 77)

The poem figures an ongoing political responsibility to Northern Ireland (echoed in the "untroubled dust" that evokes the Troubles) as an aesthetic responsibility.

Sitting beneath a map, the lyric "I" of this poem is at once subject to the laws of history and geography and autonomous, perhaps capable through art of achieving the levitational power available only beyond the laws of Earth—for an astronaut, for Christ, or, best yet, for Christ on the moon. Heaney's lyric otherworlds navigate the internal and the external to create a lyric "I" responsive to and embedded in history but not completely determined by it. Heaney uses astronomy at crucial moments in his oeuvre, most often when he is being asked to perform a public, collective role, to imagine a more ethical and autonomous lyric world into being.

Heaney develops lyric otherworlds in part through his myth of the bog. The *ars poetica* "Digging" declares that Heaney's work will come out of the peat bogs of his childhood landscapes. Due to their low oxygenation, the bogs preserve anything that disappears into them in pristine shape and offer an archive of natural and ancient human history where butter one thousand years old and a great elk may exist together. Heaney developed the myth of the bog most fully in *North* (1976), but it begins in his early volume *Door into the Dark* (1969) as an alternative to the myth of the prairie and westward expansion that suffused the race to the moon. Against American

---

[76] Robert Tracy remarks on this title change in Tracy, "Westering: Seamus Heaney's Berkeley Year," *California Magazine*, 10 Jan. 2018, https://alumni.berkeley.edu/california-magazine/online/westering-seamus-heaneys-berkeley-year/.

individualism, Heaney's bog poems frequently adopt a first-person plural perspective in responding to the Troubles. In the late 1960s, he became preoccupied with the eerily preserved Iron Age bog bodies found in bogs of the Jutland Peninsula that many anthropologists presumed were sacrificed to the earth goddess. Inspired by Danish archaeologist P. V. Glob's *The Bog People* (1969), a volume filled with photographs of Iron Age bodies preserved in Scandinavian peat bogs, Heaney's poems compare the ritualistic killings of the Troubles with the human sacrifices of Iron Age northern Europe.

The very earthiness of the bog myth imbues it with otherworldliness. Reading Heaney's bog poems, Sumita Chakraborty finds that these ecological poems are also extraterrestrial ones. She links the fetal imagery that surrounds Heaney's unearthed bog bodies to photographs of planet Earth taken from space. Heaney's work straddles a move from the uncanny as woman's body to the planetary uncanny that Gayatri Spivak charts at mid-century, leading Chakraborty to argue that "Heaney's later work stakes a claim for the uncanny as a fundamentally poetic concern."[77] Chakraborty's insights about Heaney's uncanny extraterrestrial lyric poems illuminate the ways in which his work might be considered "cosmopolitan." The bog myth itself is a time-traveling cosmopolitanism, linking the living and the dead of the ancient past in a repository of communal history. This cosmopolitanism is rooted in difference and encounters with otherness. Often, bog bodies are described not only in fetal but also in alien terms. In "The Tollund Man," based on the bog body of that name in Glob's book, Heaney's speaker finds kinship with this state of alienation. The poem has the fantastical lyric temporality of many of Miłosz's poems. It begins in the future tense and imagines a journey to a dreamy, unspecified time when the speaker will stare at an ancient body that does not look quite human:

> Some day I will go to Aarhus
> To see his peat-brown head,
> The mild pods of his eyelids,
> His pointed skin cap.

(OG, 62)

The poet feels a cosmopolitan kinship with the Tollund Man through their shared experiences of ritual killings in northern Europe:

---

[77] Sumita Chakraborty, "Of New Calligraphy: Seamus Heaney, Planetarity, and Lyric's Uncanny Space-Walk," *Cultural Critique* 104 (Summer 2019): 101–34, here 102.

> Out there in Jutland
> In the old man-killing parishes
> I will feel lost,
> Unhappy and at home.
>
> (63)

Heaney often compares travels to other lands to space voyages, from the moon landing of "Westering" to the "space-walk" of Stockholm. The imagined visit to Aarhus has a similar quality of alterity.

North (1976), the controversial Troubles volume in which Heaney develops the bog myth, teems with astronomical images that capture not cosmopolitan hope but, rather, cosmic despair. Northern Irish astrophysicist Jocelyn Bell Burnell, to whom Adrienne Rich refers in "Planetarium" (see Chapter 2), includes two poems from Heaney's Troubles period in her anthology *Dark Matter: Poems of Space*.[78] One of these is "Westering"; the other is "Exposure," from "Whatever You Say, Say Nothing" in *North*. The poem is set in Wicklow, in the Republic of Ireland, where Heaney has "escaped from the massacre" of the North but is haunted by others' criticisms for the move. Again using a metaphor of gravity and scale, he writes of "weighing" his responsibility to the crisis, even making a self-consciously inflated comparison of himself to Mandelstam, imprisoned for his poetry written in Stalinist Russia. Ultimately, he suggests he is no visionary and remarks on his poetry's inability to offer cosmic salvation to the North:

> A comet that was lost
> Should be visible at sunset,
> Those million tons of light
> Like a glimmer of haws and rose-hips,
>
> And I sometimes see a falling star.
> If I could come on meteorite!
>
> (*OG*, 135)

He goes on to liken himself, wryly, to the comet: he is an "inner émigré, grown long-haired." Heaney plays with the traveling etymology of "comet" to suggest cosmopolitanism and the vast distances of the émigré. The word moved from the Greek κομήτης ("wearing long hair," or "long-haired star") to the Latin *comēta* to the Old English *cometa*. But he also builds himself into a specifically

---

[78] Maurice Riordan and Jocelyn Bell Burnell, *DarkMatter: Poems of Space* (Lisbon: Calouste Gulbenkian Foundation, 2008), 117–18 ("Exposure"); 208–9 ("Westering").

Irish émigré tradition here, evoking the comet near the end of James Joyce's *Ulysses* (itself a work that is self-consciously cosmopolitan). The "Ithaca" homecoming chapter of *Ulysses*, in which Leopold Bloom returns to his wife after his day in Dublin, uses the comet as its operative metaphor, an image of motion and return in its epic cosmic orbit. The comet frames Heaney's poem, returning at the end when the poet worries that he has "missed / The once-in-a-lifetime portent, / The comet's pulsing rose" (136). The rose is a symbol of an invented heroic Ireland, the dark *roisin* featured in W. B. Yeats's Irish revolutionary poetry. "Rose" evokes both the flower and the idea of rising, as in a rebellion or revolution, contrasted with the "falling star" earlier in the poem. "The comet's pulsing rose" is an image of rooted cosmopolitanism and also of a political revolution that the poet—and indeed, poetry itself—might not take part in.

The retrospective poem "The Flight Path" (1996) returns to the ostensible "betrayal" Heaney committed in moving away from Northern Ireland, captured in the title underscoring air travel. "The Flight Path" dramatizes an encounter with Sinn Fein spokesman Danny Morrison on a train at the beginning of the hunger strikes in 1981. Sinn Fein, often translated as "ourselves alone," has been notoriously group-oriented, and Heaney's anecdote asserts the triumph of lyric integrity over group coercion. Entering Heaney's train car, Morrison asks, "When, for fuck's sake, are you going to write / Something for us?" Heaney replies, "If I do write something, / Whatever it is, I'll be writing for myself."[79] He suggests that the alternative to this coercion and weight is the free-falling American scene he also alludes to in "Westering": "Up and away. The buzz from duty free" and then "The spacewalk of Manhattan. The re-entry" (*SL*, 23). The poem goes on to describe global travel through the conceit that it takes light a long time to travel across interstellar distances, evoking the disconnect the "inner émigré" always feels between home and away (25). It ends on a cliff in France with a lizard whose legs resemble "a moon vehicle," a comparison that opens into the poem's concluding image of a dove rising above the landscape (26). Having missed "the comet's pulsing rose" and the "falling star" in the public poem "Exposure," here the poet finds a moment of peace in lyrical contemplation.

Heaney wrote "The Flight Path" in the 1990s, a decade marked in Ireland by the peace process and an optimistic view of the possibility of a multicultural democracy. In U.S. foreign policy of this period, Northern Ireland was often described as a model for other places around the world in which the Euro-American empire was trying to keep "peace," from Bosnia to Rwanda to

---

[79] Seamus Heaney, *The Spirit Level* (New York: Farrar, Straus and Giroux, 1996), 25. Further references are noted in the text as *SL*.

Israel. Heaney's work of the 1990s, including *The Cure at Troy* that Bill Clinton quoted from on his visit to Derry, reflects this climate. Yet the poetry still suggests its fantastical properties, its deliberate distance from the world as it is. "Lightenings viii," from the volume suggestively titled *Seeing Things*, describes a flying ship as an allegory for intercommunal cooperation that was not coming to pass in Northern Ireland. Heaney transforms a medieval story from Ireland into a story about two sides, from two opposing realities, who come together to help one another without question, as the monks of Clonmacnoise monastery free a flying ship whose anchor gets stuck in their altar (*OG*, 388). The poem evokes the Celtic otherworld, reached through a ship in a magical voyage called the *immram*, mixing Celtic mythic and Catholic tropes. In many accounts, the hero is unable to return from the otherworld; here, his ship is free to sail on through a basic act of help that only those inhabiting another reality are able to offer. The ship that defies gravity as it sails in an otherworld becomes an image of peace to come in a marvelous dream of cooperation, one that in 1991 seemed as fantastical as a flying ship. The flying ship Heaney pulled from medieval annals has a similar function to Miłosz's elegiac unicorn, a creature that relegates a world that could believe in hope to the past. In his space-themed Nobel speech, Heaney would again discuss the lyric's ability to defy gravity. After remarking that "poetry" made "this space-walk possible," Heaney adds, "Poetic form is both the ship and the anchor. It is at once a buoyancy and a holding," a form "true to the impact of external reality and... sensitive to the inner laws of the poet's being'" (*SL*, 430, 420). The lyric is an intermediary force, the anchor between this world and another, that offers encounters with otherwise unimaginable alterity, encounters that can reimagine the world so completely that ships could fly.

Following the Irish Republican Army ceasefire in 1994, Heaney wrote an essay that again evokes gravitation: "Light finally enters the black hole."[80] A black hole is an object so massive that not even light can escape; the peace process evokes, in contrast, "loosening gravity." Heaney frequently compares the work of the lyric to finding the right tug of gravity and buoyancy, attention to material conditions and the fantastic possibility of something beyond. "The soul at anchor" balances ascent into the marvelous in "A Kite for Michael and Christopher"; in "Old Smoothing Iron," domestic and poetic work are likened to "a tug at anchor," to "Feel dragged upon. And buoyant" (*SL*, 215, 206). Osip Mandelstam, Heaney reflected near the end of his life, reminded him during the Troubles "that the anchor of poetry had to be lifted off the bottom of your

---

[80] Seamus Heaney, "Light Finally Enters the Black Hole," *Sunday Tribune*, 4 Sept. 1994, A9.

ear and should drag a certain amount of your inwardness up along with it" (*SS*, 175). Heaney's final work, the posthumously published translation of *Aeneid Book VI*, teems with references to astronomy and gravity. This book of the *Aeneid* is prophetic: in a parallel of the Orphic lyric journey, Aeneas visits his father in the underworld and learns of the atrocities, written into the stars, that will occur in the founding of Rome. Anchises prophesies that Caesar will extend his empire into the stars but withholds knowledge of the suffering and violence to come en route to this cosmic glory.[81] Heaney translates the final line of the book into an image of lyric stasis and hope planted at the end of an epic that predicts bloody conquests and grisly futures, as the boats are cast off their anchors before sailing into bloody conquests.[82]

"Alphabets" anticipates Heaney's final turn from lyric to epic, the trajectory of the Virgilian poetic career. This mid-career poem is, in fact, Heaney's definitive work on the poetic vocation; it is also the major world-creation poem of his career. Here he traces a lyric cosmology, from the tiny scale of the ovum to the vast scale of the planet, bound together in the evocation of lyric address that links all of time and space. Several critics have approached the poem in both global and planetary terms, but not quite in world-making ones. Ramazani argues that "The poem is about the globe and is itself a globe." He goes on,

> The binding force of the poem's sonic and structural form represents the globe in uniquely poetic fashion, neither as an economic system of equivalence nor as an imperial domain but as an auditory and visual artifact, an imaginative structure, that binds together and sets free meanings, resonances, analogies, languages, and patterns.[83]

Ramazani describes more an act of worlding than a globe as conventionally understood. For Chakraborty, the poem is more of a planet than a globe. In her reading, the globe transforms from an imperial artifact and into an uncanny planet seen from space, one that Heaney links to the uncanny apostrophic "O". Chakraborty argues that Heaney foregrounds the apostrophic "O" to "show us its strangeness," turning it into Spivak's uncanny O of planet Earth by employing it indirectly and suspending it at the end of the line: "In splitting off the 'O' from the structure of addressor and addressee, he has given us the ability to see its strangeness," allowing him to "draw attention to lyric as an

---

[81] Seamus Heaney, trans., *Aeneid Book VI* (New York: Farrar, Straus and Giroux, 2016), 83, 89.
[82] Heaney, *Aeneid Book VI*, 93.
[83] Jahan Ramazani, "Seamus Heaney's Globe," *The Irish Review* 49/50 (Winter–Spring 2014/15): 38–53, here 42.

improbable object of its own."[84] Through apostrophe, then, Heaney suggests the strangeness of lyric reading and the ethical potential of that strangeness to imagine modes of collective belonging not predicated on the imperialist logic of domination through epistemological mastering of the universe and the colonial other. But beyond its likening lyric address to a globe or a planet, "Alphabets" posits the lyric poem as an otherworld of its own.

"Alphabets" opens *The Haw Lantern* (1987), Heaney's volume with the most obvious Eastern European influences.[85] The best-known poem from the volume is "From the Republic of Conscience," written for Amnesty International and from which the Amnesty International Award takes its name; in it, the poet lands in the "Republic of Conscience," where he becomes a dual citizen and a lifetime ambassador (*OG*, 277). "Alphabets" should be understood in the tradition of the cosmopolitan émigré, the Irish *filidh* tradition, and Eastern European parable verse. Heaney notes that the poem was influenced by Miłosz's "The World (A Naïve Poem)," a text that imagines, in childlike and ecstatic terms, the world as the poem claims it should be. Heaney measures the distance of this imagined world from the one we inhabit in part through the motif of translation, and not only between English and Latin. The motif of translation includes Miłosz, whose work Heaney could not read in the original Polish. "Alphabets" describes learning to write in Latin as "new calligraphy that felt like home," but calligraphy is also a word Heaney associated with Miłosz. Elsewhere, he describes "the obstinate espousal of beauty and order—the calligraphy, as [Miłosz] once called it—of poems like 'The World.'" (*SS*, 281). The line is a nod to Miłosz as an inventor of lyric worlds, and a translation of the world we inhabit into a lyrical one. "Alphabets" also adopts Miłosz's sense of poetry as incantation. In one of its incarnations, the poem's O is the magical figure of a necromancer, who kept "[a] figure of the world with colours in it" (*OG*, 271). Heaney refers here, as Neil Corcoran notes, "to the Renaissance Neoplatonist Marsilio Ficino as he is described in Frances Yates's *Giordano Bruno and the Hermetic Tradition*. Ficino's 'figure of the world' is a magical talisman designed to gain benefit from the universe."[86] In fact, Corcoran notes, "Alphabets" quotes Yates's account of Ficino almost verbatim. This account gives the poet the power to "'[arrange] the figure of the world with knowledge and skill."[87] By bringing this allusion into the poem, Heaney casts the poet, in the *filídh*-Yeatsian

---

[84] Chakraborty, "Of New Calligraphy," 128.
[85] Heaney discusses the Eastern European poetic parable quality of *The Haw Lantern* in *SS*, 292–3.
[86] Neil Corcoran, *Poets of Modern Ireland: Text, Context, Intertext* (Carbondale: University of Southern Illinois Press, 1999), 79–84, here 81.
[87] Corcoran, *Poets of Modern Ireland*, 82.

tradition, as a magician who can conjure a harmonious world through the skills of poetic craftmanship. This magical bringing into being of a world (linked to *poiesis*, or making) also speaks to the *filídh*, the highest representatives of the *áes dána* group, meaning "people of skill, craft." The magical globe, the medieval figure of the universe, Giordano Bruno's doctrine of the plurality of worlds (also in Miłosz's "Campo dei Fiori"), and the planet suspended in outer space come together in the lyric's crafting of an otherworld.

Heaney and Miłosz share nothing less than a method of imagining the poetic vocation as a means of world-making. In this sense, Miłosz's cosmic theology chimes with Brodsky's sense of the Orphic lyric as an origin story for both the universe and the lyric, suggesting a cosmological and cosmopolitan poetic heritage. It is fitting, then, that Heaney's elegy and prose tribute to Miłosz, both of which take stock of his career and influence, are astronomical. The elegy "Out of This World" (2004) draws on the poets' shared backgrounds as European Catholics, beginning with an elegiac catechism—"Q. *Do you renounce the world? / A. I do renounce it.*"—and blending the zaniness of the title "out of this world" with the seriousness of Catholic mysticism.[88] In the elegiac tribute "In Gratitude for All the Gifts," Heaney ties Miłosz's poetic craft and vocation to the intimate cosmic relationship between words and worlds: "His life and works were founded upon faith in 'A word wakened by lips that perish.'" Heaney quotes from the last third of the poem "Meaning," an agnostic and cosmic self-elegy in which Miłosz imagines not what lies beyond the world, but the world's composition: "When I die," Miłosz writes, "I will see the lining of the world." This convergence of *lining* and poetic lineation, of *world* and *word*, is constitutive of what Heaney deems Miłosz's "first artistic principle...clearly related to the last gospel of the Mass, the In principio of St John: 'In the beginning was the Word.' Inexorably then," Heaney continues, "through his pursuit of poetic vocation...he developed a fierce conviction about the holy force of his art, how poetry was called upon to combat death and nothingness." Following the declarative opening lines of "Meaning" is a series of questions, doubts about what there will be to "see": "And if there is no lining to this world?" the speaker asks, "And on this earth there is nothing except this earth?" Miłosz's answer to these doubts is poetry itself, the "word wakened" that collapses into meaning with its poet progenitor to, as Heaney puts it, "combat death and nothingness" as "A tireless messenger who runs and runs / Through interstellar fields, through revolving galaxies, / And calls out, protests, screams."[89]

---

[88] Seamus Heaney, *District and Circle* (London: Faber & Faber, 2011), 48.

[89] Heaney, "In Gratitude for All the Gifts," *The Guardian*, 10 Sept. 2004, https://www.theguardian.com/books/2004/sep/11/featuresreviews.guardianreview25; Miłosz, "Meaning," in *New and Collected Poems*, 569.

5

"Vast and Unreadable"

Tracy K. Smith, Astronomy, and Lyric Opacity in African American Poetry

Speaking at a Stanford University event on poetry and astronomy, a screen displaying colorful nebulae and galaxies behind her and a predominately white audience in front of her, Tracy K. Smith described the influence of Hubble Space Telescope images on her poetry. These stunning images form the backdrop of her volume *Life on Mars* (2011), an elegy for the poet's father, William Floyd Smith, one of the engineers who worked on Hubble.[1] In homage to his legacy, Smith notes that she struggled to pull her mind back to the "facts" of astrophysics when she found her imagination distracted by the "painterly" and "curated" quality of the photographs. Yet Hubble images are not unmediated scientific data; they invoke both sublime aesthetics and sublime science. Elizabeth Kessler demonstrates how astronomers adjust color, contrast, and composition of Hubble images to "encourage a particular way of seeing the cosmos" through "the visual language of the sublime." In doing so, Kessler argues, "they encourage the viewer to experience the cosmos visually *and* rationally, to see the universe as simultaneously beyond humanity's grasp and within reach of our systems of knowledge."[2] Burke and Kant locate the sublime in intense aesthetic experiences that threaten to overwhelm subjectivity, often through scales that exceed human comprehension. They link sublime experiences both to starry skies and to the bodies of racial others, which astonish the implicitly white subject with their unfathomability.[3]

---

[1] Tracy K. Smith, "Imagining the Universe: *Life on Mars* with Tracy K. Smith," Stanford University, 1 Dec. 2014, http://www.youtube.com/watch?v=QW7WIJaSfl4&t=110s.

[2] Elizabeth A. Kessler, *Picturing the Cosmos: Hubble Space Telescope Images and the Astronomical Sublime* (Minneapolis: University of Minnesota Press, 2012), 3.

[3] In Burke's account of the sublime, a white woman might inspire beauty, while a Black woman inspires the horror associated with the sublime, as he recounts in "the story of a boy, who had been born blind," who saw a Black woman for the first time and "was struck with great horror" (Burke, *A Philosophical Enquiry into the Origin of Our Ideas of the Sublime and Beautiful*, 116). For an account of Burke's and Kant's gendered and racialized sublime aesthetics, see Christine Battersby, *The Sublime, Terror, and Human Difference* (London: Routledge, 2007), especially 11–12.

In the sublime tradition, then, both the cosmos and blackness are analogues for unknowability.

Life on Mars reroutes the sublime into strategies of opacity that protect the privacy of its Black subjects. The real and imagined people of *Life on Mars*, including the poems' speakers, always remain a little unfathomable, just out of reach of one another and of their readers. The sublime Hubble metaphor brings extreme scales and visual aesthetics into a volume already predicated on a scalar contrast: Smith elegizes her father through cosmic metaphors, inflating his loss to the size of the universe and personalizing the cosmos to match the homey and private tenor of his singular loss. The volume teems with collectives marked by the pronoun "we"—a species "we" held together on a vanishingly tiny planet, a collective Black "we" that never quite emerges— but *Life on Mars* ultimately explores collective relations through the small-scale, irreducible lyric "I" and its relationship to the volume's other insistent pronoun, "it." *It* does not speak to dehumanization so much as to the limits of what the "I," the "you," and the "we" can know about one another. The ambiguous pronoun circulates throughout the volume, chased by similes that never pin the opaque cosmic concept down. As Smith describes in "It & Co.," "It is like some books, vast and unreadable."[4] By drawing on opacity at the level of pronouns, visual idioms, and the sublime, Smith channels Hubble images into strategies of what I will call "lyric opacity" to develop an anti-racist planetary ethics that does not lose sight of the personal.

At the end of "My God, It's Full of Stars," Smith recalls Hubble's initial optical glitch and its subsequent repair as a metaphor for the limits of knowledge:

> The first few pictures came back blurred, and I felt ashamed
> For all the cheerful engineers, my father and his tribe. The
>   second time,
> The optics jibed. We saw to the edge of all there is—
> So brutal and alive it seemed to comprehend us back.
>
> (*LOM*, 12)

When Hubble's "optics jibed" following repairs, it sent back startlingly clear images, many of which seem to "watch" us because they resemble eyes, such as the Cat's-Eye Nebula (NGC 6543), the Eye-Shaped Planetary Nebula (NGC 6826), and the Glowing Eye (NGC 6751). Smith elegizes her father throughout the volume through optical tropes, which mark the limits of knowledge.

---

[4] Tracy K. Smith, *Life on Mars* (Minneapolis, MN: Graywolf, 2011), 17. Further references are noted in the text as *LOM*.

Here, the vague "it" of the cosmos only "*seemed* to comprehend us," calling attention to the limits of knowledge the volume explores—of the self, the family, the past, the afterlife, and the cosmos. The imprecise "edge of all there is" only "seemed" to look back. That essential "seemed" emphasizes limits in knowledge and relations. Of course, the "it" does not really "comprehend us," any more than we, "with our faulty eyes," can read "it" (14). The word "seemed" loops us back to the poem's playfully abstract opening line: "We like to think of it as parallel to what we know" (7). "Like," "as," and "parallel" all evoke simile without actually employing it; they are a simile for a simile, a device that holds open and qualifies the space a metaphor would purport to close. At best, we get something *like* knowledge of a cosmic "it." The volume has an archival impulse to understand and document William Floyd Smith's past, much of which is as unfathomable to his daughter as his place in the afterlife. In its combination of elegy and the extraterrestrial, the volume amplifies the effects of opacity.

As the concluding critical intervention of this book, I offer "lyric opacity" as a term for a kind of poem for which we do not have a critical idiom: one whose language is accessible and inviting, even "transparent," but whose subjectivity is slippery and unfathomable as the poem navigates the politicized, racialized, and gendered expectations of the lyric "I." In what follows, I expound on lyric opacity through the verse of Smith and other poets alongside whom Smith's work should be considered. These poets include Rita Dove and Natasha Trethewey as well as a cluster of Smith's self-proclaimed "most necessary poets," including Seamus Heaney and Yusef Komunyakaa.[5] It might strike readers as odd to characterize the eminently approachable work of the poets I describe here as "opaque." I don't use this term in Viktor Shklovsky's sense of "difficult" poetry, with its "roughened, impeded language" meant to slow down reading.[6] Shklovsky's version of opaque, closely related to *ostranenie* (or Brechtian estrangement), has influenced avant-garde poetics. Ron Silliman, in his influential blog on contemporary poetry, puts this Shklovsky–Brechtian lineage of poetry in opposition to what he terms the "School of Quietude"—poetry invested in traditional structures that considers itself pure, unmediated poetry, "poetry's unmarked case."[7] Contra Silliman, Smith's "quiet" poetry in fact inclines toward estrangement—not at the level of

---

[5] Tracy K. Smith, *Ordinary Light* (New York: Vintage, 2015), 336. Further references are noted in the text as *OL*.
[6] Viktor Shklovsky, "Art as Technique" in *Literary Theory: An Anthology*, ed. Julie Rockin and Michael Ryan (Oxford, UK: Blackwell, 2004), 15–21, here 16.
[7] Ron Silliman, "Wednesday, July 7, 2010," in Silliman's Blog, 7 July 2010, http://www.ronsilliman.blogspot.com/2010/07/i-know-whenever-i-use-phrase-school-of.html.

difficult language but, rather, at the levels of genre and subjectivity. Smith's experiments with genre bring the transparency of language and the inapprehensible cosmic backdrop into chilling tension in ways that make her lyric "I"s and their relations with others strange and opaque. This kind of opacity—of protecting the self and others from a demand for full exposure—is quietly oppositional in the contemporary U.S. political climate. Smith was named Poet Laureate under the Trump Administration, which provoked highly visible and explicit forms of opposition, including the "exposure" (an optical metaphor) of government incompetence and conspiracies. Christina León has done remarkable work on artists who develop an "opaque aesthetics" to respond to this cultural moment, artists who induce audiences "to get stuck with [the] work, tarrying with its opacity and abstraction, rather than gliding over nuance and flattening the work as reducible to sociopolitical content."[8] While the effect of Smith's opacity is often to keep her work from being reduced to a sociopolitical position, I do not employ "opacity" quite in León's Shklovsky-inflected sense. Smith's opacity emerges not in language or form but rather in voice, that poetic figure often identified with the lyric "I."

Édouard Glissant offers an account of opacity in an anti-colonial, anti-racist context that is more amenable to quiet aesthetics. Glissant's opacity describes an ethical mandate to respect the privacy and unknowability of the other. In his early work, Glissant defines opacity as an aesthetic, ethical, and political strategy to counter globalization's insistence on transparency, which seeks to expose the "other" so that differences across cultures can be assimilated into a Eurocentric notion of a common humanity.[9] Glissant's later writings expand the scope of opacity beyond a postcolonial Caribbean context to a broad strategy of ethical relations. In a late interview, Glissant candidly summarized the ethical implications of opacity: "[A] person has the right to be opaque. That doesn't stop me from liking that person, it doesn't stop me from working with him, hanging out with him, etc. A racist is someone who refuses what he doesn't understand. I can accept what I don't understand."[10] Glissant's focus on race as a key element of opacity touches on a fundamental

---

[8] Christina León, "Forms of Opacity: Roaches, Blood, and Being Stuck in Xandra Ibarra's Corpus," *ASAP/Journal* 2:2 (May 2017): 369–92, here 372.

[9] See especially Édouard Glissant, *Caribbean Discourse: Selected Essays*, trans. J. Michael Dash, CARAF Books (Charlottesville: University of Virginia Press, 1989), 118; and *Poetics of Relation*, trans. Betsy Wing (Ann Arbor: University of Michigan Press, 1997), 111–20.

[10] Édouard Glissant, "Conversation with Édouard Glissant Aboard the Queen Mary II," interview by Manthia Diawara, trans. Christopher Winks, in *Afro-Modern: Journeys Through the Black Atlantic*, ed. Tanya Barton and Peter Gorschlüter (London: Tate, 2010), 58–63, here 62.

concern in optical physics: degrees of visibility. Opacity is also a specifically poetic strategy for Glissant, channeling obscurity into moments of startling clarity that he renders in visual terms; "it is a flash of lightning that hesitates and overwhelms: poetry."[11] Opacity, then, does not suggest willful vagueness but, rather, a controlled and intentional vacillation between withholding and revealing that allows for more ethical encounters with others, be they at a micro level (the "I" and "thou" often staged in a poem, for example) or at a macro level (collective relationships operating on national, global, or planetary scales). It is in this ethical, relational thrust that the term "opacity" carries a different meaning than other, related words with heft and presence in African American literature, such as "invisibility." While opacity often does address and redress how the white gaze renders Black bodies and contributions invisible, it does so specifically through investigating the ethics and aesthetics of knowledge. A lyric voice that refuses to fade into, align with, or elucidate a collective historical or sociopolitical issue—a voice, in other words, that insists on its singularity through upholding its right to privacy or incomprehensibility—is what I refer to here as opaque lyric subjectivity.

Smith's work, like Glissant's, is invested in the science of opacity, which is, first, an optical term referring to how much light can pass through a given medium that reflects, absorbs, or scatters varying amounts of it. Opacity is a metaphorically rich idea for a volume with a literal lens, the Hubble Space Telescope, as its organizing figure. The giant "eye" of Hubble has brought back many poignantly clear images of cosmic bodies that only throw into relief how little we know about the cosmos. The clarity of these images belies their very mysteriousness: the better we can see them, the more we witness their unfathomability—just as we do not know what the dead are, or where they go. This clarity also lends itself to interpretive opacity by bringing into exquisite detail everything that the poems cannot know.

5.1 Astronomy and African American Poetry

Life on Mars continues a long tradition of African American astronomical poetry, as well as discourses of science fiction and Afrofuturism tied to expressions of personhood. Phillis Wheatley's early work in this tradition draws on Burke's and Kant's theories of the sublime just coming into popular

---

[11] Édouard Glissant, "The Thinking of the Opacity of the World," trans. Frank Loric, *Frieze d/e* 7 (Winter 2012): 77.

circulation at her time of writing. Her poem "On Imagination" evokes the sublime to suggest that the act of stargazing unleashes the immeasurable soul:

> We on thy pinions can surpass the wind,
> And leave the rolling universe behind:
> From star to star the mental optics rove,
> Measure the skies, and range the realms above.
> There in one view we grasp the mighty whole,
> Or with new worlds amaze th' unbounded soul.[12]

The form of the poem mimics cosmic harmony. The neat rhymed couplets in iambic pentameter themselves "measure the skies," a pun on poetic meter. But these realms, like the soul, are ultimately immeasurable. The references to stellar navigation, and perhaps also the "rolling universe," evoke the churning seas of the Middle Passage, across which a slave ship took Wheatley at age seven or eight following her kidnapping in Senegambia and which she again crossed in 1773 from Boston to London in pursuit of publishing her poetry. In its use of the extraterrestrial to evoke unfathomable personal depths, Wheatley's "On Imagination" begins an African American tradition that looks to astronomy to convey Black subjectivity that exceeds measurement and knowability.

In the nineteenth century, stars and constellations would become vital to the African American fight against slavery, as enslaved peoples fled north through stellar navigation, using the Big Dipper to locate the North Star. The importance of this metaphor is recorded in the anti-slavery newspaper *The North Star* (1847–51), published by Frederick Douglass and briefly co-edited with fellow abolitionist and astronomy enthusiast Martin Delany. The connection between astronomy and emancipation inflected Delany's contributions to nineteenth-century science. In *Fugitive Science*, Britt Rusert describes how Delany turned to astronomy to connect "natural science and the struggle for emancipation" as he "actively linked scientific revolution to race revolution."[13] Rusert compares Delany's work on electricity and the movement of planetary bodies to his novel *Blake: or, the Huts of America*, which "transforms the archetypal fugitive slave narrative into speculative fiction." Rusert argues, "Delany's serial novel and writings on astronomy work together to show

---

[12] Phillis Wheatley, "On IMAGINATION," in *Phillis Wheatley: Complete Writings*, ed. Vincent Carretta (New York: Penguin, 2001), 36.

[13] Britt Rusert, *Fugitive Science: Empiricism and Freedom in Early African American Culture* (New York: New York University Press, 2017), 149, 152.

readers how the mobilization of extra-terrestrial metaphysics and speculative science might help forge a practical science of emancipation here on earth."[14] Rusert's astronomical emancipation records an emerging Black nationalism that Rusert, credited with the pan-African slogan "Africa for Africans," endorsed. National powers frequently write their destinies into the stars in a doctrine of "as above, so below"; we might think of the moons, suns, and stars that adorn national flags, including that of the U.S. Delany, too, invokes astronomy to imagine an alternative transnationalism: a network of fugitive cosmic bodies that defy "the circumspect boundaries placed around the human in both antebellum science and literature."[15] In this cosmos teeming with fantastical bodies of comets and stars, Delany envisions a vibrant transnational Black nation full of agential subjects.[16]

Amiri Baraka's space age poem "Planetary Exchange" also evokes an agential cosmos in the context of Black nationalism. Baraka begins with a collective "we" of mere inanimate matter before claiming an "I" in harmony with a materialist cosmos. "*We are meat in the air. Flying into night space*" gives way to a stanza of just two words: "*I am.*" After this assertion of selfhood, the "I" becomes part of a teeming, unpredictable cosmos through poetic song: "*I am. I am. I am. Through the dazzling / lives of the planets and stars. I am. sings.*"[17] Baraka channels the Whitmanian cosmic lyric "I" but challenges its universalizing purview. "I am" asserts selfhood against a white supremacist gaze and poetic tradition that has denied the humanity of the Black speaker. But this "I" does not simply move into the position of an imperial "I"; instead, it imagines the human as an interplanetary creature enmeshed with the "dazzling / lives" of cosmic bodies that, like the Black body under the white gaze, have been imagined as passive objects against which to imagine white subjectivity. Baraka ends the poem by absenting the lyric "I," replacing it with pure lyric voicing that seems to come not from the speaker ("*sings*" does not agree with "*I*") but rather from the cosmos.

Delany's *Blake* and Baraka's "Planetary Exchange" are both part of an Afrofuturist tradition. Mark Dery coined the term "Afrofuturism" to describe "[s]peculative fiction that treats African-American themes and addresses

---

[14] Rusert adds, "Delany explores how fugitive bodies, which exceed the restrictive boundaries of the human, become vectors of force and affect change in a world that stretches beyond the South and the nation-state and reaches across the cosmos" (164).

[15] Rusert, *Fugitive Science*, 164.

[16] Rusert discusses Delany's *Blake* in terms of Jane Bennett's "vibrant materiality" to demonstrate how "objects," including enslaved peoples, can "act as positive agents" (*Fugitive Science*, 173).

[17] Amiri Baraka, "Planetary Exchange," in *Black Magic: Sabotage, Target Study, Black Art: Collected Poetry, 1961–1967* (Indianapolis, IN: Bobbs-Merrill, 1969), 224.

African-American concerns in the context of twentieth-century technoculture."[18] Noting the prevalence of alien tropes in Afrofuturism work, Dery writes, "African Americans, in a very real sense, are the descendants of alien abductees."[19] For Toni Morrison, in an interview with Paul Gilroy, this sudden and violent dislocation from homeland makes enslaved Africans the first modern subjects.[20] Robert Hayden writes from a literally alien perspective in his "[American Journal]" (1978), told from the perspective of an extraterrestrial visitor taking notes on his visit to America to type up in a report to "The Counselors." The final poem in his *Collected Poems*, "[American Journal]" takes stock of Hayden's poetic legacy. It is a persona poem, evoking Du Bois's "double consciousness" in the dividing of the voice. For Du Bois, double consciousness is not so much a true "doubling" as a fracturing, as the Black subject comes to see themselves through dominant white structures. The Black subject, Du Bois argues, is "born with a veil, and gifted with second-sight in this American world,—a world which yields him no true self-consciousness, but only lets him see himself through the revelation of the other world."[21] Hayden experiences his own poetic voice as, in part, alien. He identifies the situation of the Black poet in America with the challenges of being an observing alien invader; he wonders "how best...disguise myself in order to study them unobserved." Turning primitivist rhetoric against the European descendants who cast indigenous peoples and Africans as subhuman, Hayden's alien speaker provincializes American exceptionalism:

> charming savages   enlightened primitives   brash
> new comers lately sprung up in our galaxy   how
> describe them   do they indeed know what or who
> they are   do not seem to   yet no other beings
> in the universe make more extravagant claims
> for their importance and identity

Even as he critiques "the Americans," the speaker confesses that he is

---

[18] Mark Dery, "Black to the Future: Interviews with Samuel R. Delany, Greg Tate, and Tricia Rose," in *Flame Wars: The Discourse of Cyberculture*, ed. Mark Dery (Durham, NC: Duke University Press, 1994), 179–222, here 180.
[19] Dery, "Black to the Future," 180.
[20] Paul Gilroy, "Living memory: a meeting with Toni Morrison," in *Small Acts: Thoughts on the Politics of Black Cultures* (London: Serpent's Tail, 1973), 175–82, here 178.
[21] W. E. B. Du Bois, *The Souls of Black Folk* (New York: Norton, 1999), 10–11.

> curiously drawn   unmentionable   to
> the americans   doubt i could exist among them for
> long however   psychic demands far too severe
> much violence   much that repels   i am attracted
> none the less[22]

Here, Hayden experiences his poetic voice as alien, and observes, as a literally alienated subject, the "America" that dehumanizes Black people. Hayden's blending of African, European, Latinx, and Bahá'í forms is often taken as evidence of his success at becoming a "poet" pursuing "the truth of human experience" rather than being a *"black poet"* preoccupied only with "the ethnocentric."[23] But the fractured form of "American Journal" captures the strain, as the predominately white culture the alien visitor observes created "the human" in the first place. As Harryette Mullen and Stephen Yenser describe, "Hayden does not so much transcend race as he employs racial identity as a metaphor for the opacity of the self."[24] The self, rendered alien by the white gaze it then sees itself through, is opaque to itself and others. As Yusef Komunyakaa puts it, "this voyage into the brutal frontier of the American experience is a confrontation with his own alienation."[25]

5.2 Lyric Opacity

Life on Mars is similarly a book about alienation approached through opacity; it contends with atrocities on Earth so terrible, from racist violence to environmental degradation, that life on Earth feels as alien as Mars. And, like Hayden, Smith approaches the topic through opacity. Smith, like Hayden, is often celebrated for writing poetry that is not "just about" race or politics, as though those formations are ever escapable. But that doesn't mean that these poems do not register and redress the predominately white history and

---

[22] Robert Hayden, *Collected Poems*, ed. Frederick Glaysher (New York: Liveright, 2013), 192.

[23] Laurence Goldstein and Robert Chrisman, eds., *Robert Hayden: Essays on the Poetry* (Ann Arbor: University of Michigan Press, 2013), 68, 16.

[24] Harryette Mullen and Stephen Yenser, "Theme and Variations of Robert Hayden's Poetry," in *Robert Hayden: Essays on the Poetry*, 233–50, here 241. Yusef Komunyakaa finds the same to be true of the poem. He reflects, "[Hayden] always saw his work as totally American. Yet I believe he identifies with the displaced speaker in '[American Journal]'—the outsider" ("Journey into '[American Journal]'," in *Robert Hayden: Essays on the Poetry*, 332–4, here 334.

[25] Yusef Komunyakaa, "Journey into '[American Journal]'," in *Robert Hayden: Essays on the Poetry*, 332–4, here 332.

formulations of lyric subjectivity. Recent work has examined the whiteness of the lyric "I." Reading confessional poetry as an outgrowth of the expressive lyric, Christopher Grobe delves into the privileges of race and gender that allowed Robert Lowell to be read as universally resonant. Grobe argues that, "whereas white women risked seeming merely private, merely personal, and merely unique, writers of color were in danger of coming across as merely public, merely social, and merely representative."[26] Thus, by Grobe's logic, a woman of color poet faces a vexing contradiction: she risks being read as an expressive "I" who is at once "merely personal" and "merely public," navigating undesirable extremes rather than striking the Lowellian harmonious balance. Kamran Javadizadeh explores how Claudia Rankine has grappled with the "white subject" of the expressive lyric, arguing that Rankine reroutes "the Lowellian investment in the singular self" to "retain the intimacy allowed by the lyric tradition" while also attending to the violent collective pasts and presents that shape American experiences.[27] Smith reflects on the apolitical and white legacy of lyric in her essay "Political Poetry Is Hot Again," where she discusses how recent U.S. poetry challenges ideas of the expressive, autonomous lyric. Smith speculates, "Perhaps America's individualism predisposed its poets toward the lyric poem, with its insistence on the primacy of a single speaker whose politics were intimate, internal, invisible." Smith marks 9/11 as a shift in "the nation's psyche" and concludes, "[t]he lyric 'I' at this very moment is not alone.... Rather, it is speaking to a large, shifting, contradictory, multivalent body that is not guaranteed to hear or even to agree. Still, the 'I' speaks. It is speaking at once from and to something like America."[28] Because "America" was built on and is sustained through racist systems, lyric poets "speaking at once from and to something like America" will grapple with this legacy. While Smith writes in a more traditionally "lyrical" style than Rankine, she also grapples with the whiteness of lyric subjectivity—especially as her work so often brings lyric to the edge of other traditionally white masculine discourses like astrophysics and science fiction that anchor scientific authority in an apparently "neutral" and "universal" (which is to say, white) perspective.

Critiques of the whiteness of the lyric "I" have not considered the role scientific authority plays in constructing and naturalizing white lyric

---

[26] Christopher Grobe, *The Art of Confession: The Performance of Self from Robert Lowell to Reality TV* (New York: New York University Press, 2017), 38–9.

[27] Kamran Javadizadeh, "The Atlantic Ocean Breaking on Our Heads: Claudia Rankine, Robert Lowell, and the Whiteness of the Lyric Subject," *PMLA* 134:3 (May 2019): 475–90, here 477.

[28] Tracy K. Smith, "Political Poetry Is Hot Again. The Poet Laureate Explores Why, and How," *New York Times*, 10 Dec. 2018, https://www.nytimes.com/2018/12/10/books/review/political-poetry.html.

subjectivity. As Peter Middleton traces in *Physics Envy*, the New Criticism emerged as a direct response to the perceived threat the epistemological authority of the sciences posed to poetry in the early twentieth century.[29] Middleton demonstrates how scientific methods and vocabulary inflected the values and methods of the New Criticism and poets who write in a New Critical vein, meaning that the sciences are in the very building blocks of the "well-made poem." Following Ezra Pound, T. S. Eliot influentially universalized a lyric "I" by imbuing it with scientific authority. In "Tradition and the Individual Talent," Eliot argues that the individual talent must extinguish itself into the larger "tradition," by which he means the implicitly white "mind of Europe." Eliot famously remarks, "It is in this depersonalization that art may be said to approach the condition of science."[30] But the ability to "depersonalize" oneself in neutral scientific terms belongs to those whose identities are reflected in a Eurocentric model of universality.

Even at her most evasive and "scientific," Smith is never impersonal. At a 2016 event celebrating Emily Dickinson and astronomy, Smith began by reading a poem seemingly remote from cosmic concerns that speaks instead to the intimacy of lyric address: "I'm Nobody! Who are you? / Are you—Nobody—too?"[31] Dickinson's poem explores lyric intimacy and relations: the "I" proclaims its hidden complexities through its professed self-negation to the readerly "you," creating a conspiratorial pair veiled from the loud, vain public world. In her memoir, Smith frames "I'm Nobody!" as an origin story for her poetic career, the first poem that worked powerfully on her: "I liked the sense of privacy the poem seemed to urge, as if there is some part of everyone—like the imagination, the spirit, or whatever gravitates toward the language of poetry—worth protecting from the world" (*OL*, 146). Smith's poems, like Dickinson's, often employ astronomical discourse to suggest the protection of personal experience. Smith disclosed at the Dickinson event that she herself is most compelled by astronomy's inward gaze, its intimacy as much as its immensity. "There are many poems of Dickinson's that are thinking up and out in physical, astronomical terms," Smith notes, but "what is happening in her work that's most chilling and comforting to me is that

---

[29] Peter Middleton, *Physics Envy: American Poetry and Science in the Cold War and After* (Chicago, IL: University of Chicago Press, 2015), 25.

[30] T. S. Eliot, "Tradition and the Individual Talent," in *The Sacred Wood: Essays on Poetry and Criticism* (1920; repr., London: Faber & Faber, 1997), 47.

[31] Tracy K. Smith and David DeVorkin, "Clockwork Universe: Emily Dickinson Birthday Tribute." This lecture took place at the Folger Shakespeare Library in Washington, D.C., on 12 Dec. 2016. The text of Dickinson's "I'm Nobody! Who are you?" (260) is taken from Emily Dickinson, *The Complete Poems of Emily Dickinson*, ed. Thomas H. Johnson (New York: Little, Brown & Co., 1976), 133.

looking into a place within the self." The juxtaposition of "chilling and comforting" describes the quality of relations explored in *Life on Mars*, relations both within the self and between the self and others. The phrase "vast and unreadable" from "It & Co." describes not only mysterious cosmic space but also inner space, the "place within the self" that is "cosmic and ultimately abstract" that Smith finds in Dickinson's astronomical work.[32] Readers glimpse an "I" that never offers full knowledge of itself or what it sees—an "I" that, as in "I'm Nobody," offers the readerly "you" material through which to explore its own vastness and unreadability.

Ironically, Smith remarked on the private dimensions of Dickinson's astronomy in a context that Dickinson would have found "dreary" and "public": the Folger Library in Washington, D.C., in December 2016, one month before the inauguration of Donald Trump. Smith concluded the event with a reading of her own "Political Poem." After winning the 2012 Pulitzer for *Life on Mars*, Smith went on to serve as U.S. Poet Laureate from 2017 to 2019, following Dove (1993–5) and Trethewey (2014–15) as the third African American woman to hold the position. (Essentially, Gwendolyn Brooks also filled the position of poet laureate from 1985 to 1986 under its former name, Consultant in Poetry to the Library of Congress. A 1985 Act of Congress changed the name to "Poet Laureate Consultant in Poetry.") Smith navigates the private and public dimensions of lyric, a task complicated by assumptions that African American poets speak publicly, collectively, and loudly rather than privately, personally, and quietly. Smith uses astronomy to investigate lyric subjectivity as her poetry reflects on and travels through racist systems in which, as Kevin Quashie puts it, "black subjectivity exists for its social and political meaningfulness rather than as a marker of the human individuality of the person who is black."[33] Smith's poetry is in a tradition Quashie identifies as the Black "quiet," with its focus on interior life through quiet (as opposed to loudly avant-garde and resistant) aesthetics. In an interview shortly after her appointment as Poet Laureate, Smith discussed poetry specifically in terms of quiet, reflecting that she hoped the public engagement dimension of the laureateship "might open up inroads to…quieter kinds of conversations." Poems, Smith explains, speak at "a decibel level that sits below the register of the media that we live with," the "advertising and the sound

---

[32] Smith and DeVorkin, "Clockwork Universe."
[33] Kevin Quashie, *The Sovereignty of Quiet: Beyond Resistance in Black Culture* (New Brunswick, NJ: Rutgers University Press, 2012), 4.

bites that we are drawn toward with and without our consent."[34] Smith's investment in poetic quiet makes her a less obviously political poet than more experimental practitioners working directly in the legacy of the Black Arts Movement.[35] However, Smith's "quietness" is itself socially determined and politically responsive. Evie Shockley has called for a more capacious understanding of what counts as "innovation" in Black aesthetics, as poets employ strategies in response to "the experience of identifying or being interpolated as 'black' in the U.S.—actively working out a poetics in the context of a racist society."[36] Shockley reminds us of the pervasiveness of racism in necessitating and activating innovations of all kinds, whether or not a Black poet writes "about" racism. The lyric subject must *always* negotiate the unmarked identity coordinates of the lyric "I"—white, male, probably straight—in ways more acutely experienced by the writer marked and read as Black.

In what follows, I will expound on lyric opacity through Smith but also through the work of some poets she asks to be read alongside. In doing so, I demonstrate how astronomy brings into focus a widespread phenomenon across contemporary poetry that works toward a decoupling of experimental form from progressive politics. Smith's poetics of opacity has precedent in two intersecting clusters of contemporary poets. The first cluster is Smith's self-designated "most necessary poets," listed in her memoir in this order: Seamus Heaney (her teacher), Elizabeth Bishop, Philip Larkin, Yusef Komunyakaa, and William Matthews (*OL*, 336). The second cluster of poets is Smith's African American women laureate predecessors, Dove and Trethewey. I will touch upon how lyric opacity functions in these works to show the concept's ubiquity and dexterity before offering a more sustained treatment of

---

[34] Tracy K. Smith, "Meet our new U.S. Poet Laureate," interview by Carolyn Kellogg, *Los Angeles Times*, 13 June 2017, http://www.latimes.com/books/jacketcopy/la-et-jc-poet-laureate-20170614-htmlstory.html.

[35] Amiri Baraka criticized this "quiet" style in a 2013 review of Charles Henry Rowell's *Angles of Ascent: A Norton Anthology of Contemporary African American Poetry*, a work he took to be "relentlessly 'anti' to one thing: the Black Arts Movement." The anthology includes Dove, Trethewey, Smith, and many other poets who either de-emphasize identity politics or engage deeply with race through inherited European lyric structures. These are poets Baraka characterizes as "university types, many co-sanctioned by the Cave Canem group, which has energized us poetry by claiming a space for Afro-American poetry, but at the same time presents a group portrait of Afro-American poets as mfa recipients." For Baraka, such institutional poets are emblematic of the anthology's post-racial celebration of pure "literary" values. Yet the lyric/politics standoff at the center of the Rowell–Baraka controversy suggests a too easy conflation of radical politics with experimental verse and respectability politics with the mannerly surface of the New Critical "well-made poem." See Amiri Baraka, "A Post-Racial Anthology?" review of *Angles of Ascent: A Norton Anthology of Contemporary African American Poetry*, ed. Charles Henry Howell, *Poetry* (May 2013), https://www.poetryfoundation.org/poetrymagazine/articles/69990/a-post-racial-anthology.

[36] Evie Shockley, *Renegade Poetics: Black Aesthetics and Formal Innovation in African American Poetry* (Iowa City: University of Iowa Press, 2011), 9.

Smith, who expands the possibilities of lyric opacity in her engagement with genre and astronomical science.

The rendering of an opaque lyric "I" through the science of opacity brings to mind above all Bishop, who, as I explored in Chapter 2, was fascinated by optical physics and astronomy.[37] Readers ubiquitously note the symbiosis of her lyric "I" and her hyper-observant "eye."[38] Bishop's "The Fish," for instance, uses optics to underscore the opacity of both the "natural" world and the human lyric subject. The speaker describes how the fish's eyes never meet her own through a careful simile: "—It was more like the tipping / of an object toward the light."[39] Bishop's intentionally vague "it" and stacked similes make their way into Smith's "My God, It's Full of Stars." As in Smith's description of stargazing humans who "toddle toward the light" (*LOM*, 7), Bishop's "object" tipped "toward the light" becomes a metaphor for seeking but not mastering knowledge. Both poems reject the possibility or even desirability of full illumination. "The Fish" ends with the entire spectrum of visible light, but not even this quasi-divine symbol marked by the order of a perfect rhyme produces epiphanic knowledge: "rainbow, rainbow, rainbow! / And I let the fish go."[40] The poem ends not by mastering the fish's otherness but rather by relinquishing the fish as an object of knowledge. The "I," like the fish, becomes stranger the more it watches; it accretes visual details without transforming those particulars into expressive revelations about the self. Bishop's "I" often slides around behind its shifting visual observations rather than defining itself through them.

Where Bishop's work investigates the science of optics, Heaney's investigates the politics of opacity. Heaney, one of Smith's teachers, achieved international fame for his political poetry written out of the sectarian divides of the Northern Ireland Troubles, culminating in his 1995 Nobel Prize. Smith is interested in Heaney's work first for its emphasis on elegy, family, and what he called the "marvellous," a sense of the divine amid atrocities that points the way toward a more ethical collective, as I explored in Chapter 4. One of only two poems Smith includes in full in *Ordinary Light* is Heaney's "Clearances" VIII, in a chapter titled after the poem. The poem is an elegiac sonnet for

---

[37] Bishop read Newton's *Opticks*, worked in an optics shop during the Second World War, and frequently alluded to optical instruments and technologies in her poetry and prose.

[38] As Mary McCarthy immortally described this effect, "Bishop creates the impression of an 'I' counting up to a hundred waiting to be found" through her "way of seeing that was like a big pocket magnifying glass." See McCarthy's "Symposium: I Would Like to Have Written....," in *Elizabeth Bishop and Her Art*, ed. Lloyd Schwartz and Sybil B. Estess (Ann Arbor: University of Michigan Press, 1983), 267–9, here 267.

[39] Elizabeth Bishop, *Poems* (New York: Farrar, Straus and Giroux, 2011), 43.

[40] Bishop, *Poems*, 44.

Heaney's mother, which Smith turned to after her own mother's death and later read at Heaney's memorial at Harvard in 2013. "Clearances" VIII begins from a different source of opacity, not a refusal to speak for a group but, rather, an inability to speak of something frighteningly illegible and almost untranslatably private. The poem elegizes Heaney's mother through a chestnut tree chopped down in the family's yard, likening the crater from the downed tree to the experience of "walking round and round a space / Utterly empty, utterly a source."[41] The empty space left by the downed tree conjures "a bright nowhere" left by his long gone "coeval / Chestnut from a jam jar in a hole," which becomes the focus of the elegy, its memory blended with his mother's into "A soul ramifying and forever / Silent, beyond silence listened for." This concluding line recalls the "church-bells beyond the stars heard" of George Herbert's "Prayer (I)," vaulting the poem's metaphysical mode into a cosmic register that evokes both the comfort and the unfathomability of the afterlife.

Heaney developed his elegiac poetics in the context of the Northern Irish Troubles, resisting the call to become a spokesperson for the Catholic community by upholding what he took to be the poet's right to privacy and autonomy, often through lyric opacity. In "Whatever You Say, Say Nothing," Heaney reflects wryly on being asked for "public" poetry on the Irish struggle: "I'm writing this just after an encounter / With an English journalist in search of 'views / On the Irish thing'" (*OG*, 123). He responds to this outside pressure to make "the Irish thing" comprehensible to the English in his follow-up poem with the optical title "Exposure," where he imagines camouflaging into a wood-kerne, "Escaped from the massacre, / Taking protective colouring / From bole and bark" (*OG*, 136). Heaney often discussed the political pressures on Northern Irish poets in terms of communal pressures on African American poets to speak on behalf of a monolithic group. In an interview the year before the publication of *Life on Mars*, Heaney reflected, "During the Civil Rights Movement in Northern Ireland we had a very strong sense of the Afro-American and Black Civil Rights Movement...and I had very strong sympathy with the Black American poets here whose first person singular, 'I,' was under pressure, always, to become first person plural, 'we.'"[42] Speaking of an earlier Irish poet and conflict, W. B. Yeats writing during the Irish War of Independence and Civil War, Robert Hayden also noted an affinity:

---

[41] Seamus Heaney, *Opened Ground: Selected Poems, 1966–1996* (New York: Farrar, Straus and Giroux, 1998), 229. Further references are noted in the text as OG.

[42] Seamus Heaney, "Making Sense of a Life," interview by Tiago Moura, *YouTube*, uploaded by *The NewsHouse*, 13 Apr. 2010, https://www.youtube.com/watch?v=s7sskc1pi_k.

I think I always wanted to be a Negro poet or a black poet or an Afro-American poet...the same way Yeats is an Irish poet....Yeats did not flinch from using materials from Irish experience, Irish myths. The whole Irish struggle has meaning for Yeats. He would have been astonished if anyone had told him to forget that he was Irish and just write his poetry. But he wrote as a poet.[43]

As in Heaney's account of the pressure on Irish and Black poets to say "we," Hayden uses the example of Yeats to suggest the desire for authorial autonomy—to use material from his experiences with racism without being personally or poetically conflated with the "Black struggle." The Northern Irish Civil Rights Movement and the Black Civil Rights Movement are, of course, historically distinct situations, the first an outgrowth of ethno-religious tensions amplified by British colonization of Ireland, and the latter of racist policies specific to U.S. national identity. The instructive commonality Heaney and Hayden isolate are the expectation for both an Irish poet and a Black American poet to be read as public and representative rather than as private and singular. Heaney's remarks on white American poetry's solipsism also reverberate in Smith's comments on poetry and politics. Heaney worked against the solipsistic tendencies of the Anglo-American lyric "I," as Chapter 4 explored. Smith, too, tempers her valuing of "the personal in poetry," with its ability to "change us," with a sense of the communal that she identifies in cultures other than white American: "I really feel that there is so much beyond the self that is and has been the subject or vehicle of poetry at other moments and in other cultures. And it's something that...American poetry has not been as willing to engage in as is necessary," she told Charles Rowell. In contrast, she proposes, "African American poets have always been conscious of a larger social context, even if they've managed to resist making it the theme or vehicle of every poem."[44] The consciousness of the social that the personal is always embedded in evokes the cultural work of astronomy, which at once delivers discoveries that concern "everyone" and offers a deeply private, contemplative pursuit of scientific observation and awe.

Like Heaney, Komunyakaa uses opacity to navigate the public and the personal. "Facing It" is an ekphrastic elegy about a literal reflective surface, the Vietnam Veterans Memorial. "My black face fades, / hiding inside the black

---

[43] Robert Hayden, "An Interview with Dennis Gendron," in *Robert Hayden*, ed. Laurence Goldstein and Robert Chrisman, 15–29, here 19.

[44] Tracy K. Smith, "'Something We Need': An Interview with Tracy K. Smith," interview by Charles H. Rowell, *Callaloo* 27:4 (Fall 2004): 858–72, here 861.

granite,"[45] the poem begins, suggesting the threatened disappearance of the singular speaker into a collective historical event—a war that disproportionately affected Black Americans, who at the time made up 11 percent of America's total population but almost 13 percent of soldiers deployed in Vietnam. The Vietnam Veterans Memorial variously functions as a wall, a window, and a mirror in the poem, sliding along a scale of opacity to transparency; a black surface that absorbs all light is fully opaque, as is a mirrored surface that reflects all light, while a window is almost fully transparent because it neither reflects nor absorbs much light. For Komunyakaa's lyric "I," transparency becomes another form of invisibility when a white veteran looks through him as they both stare at a black wall of (literally) white names: "A white vet's image floats / closer to me, then his pale eyes / look through mine. I'm a window."[46] The lyric voice teeters between concerns of overexposure and erasure as the poem tries to recuperate the individual "I" by asserting the singularity of suffering. Paradoxically, the "I" occludes itself, fading into the opaque black wall, in order to proclaim the irreducibility of the singular "I"s we cannot know, the names of the dead whose stories are lost: "In the black mirror / a woman's trying to erase names: / No, she's brushing a boy's hair." The speaker corrects his assumption about erasure into a gesture of intimate banality, but this is not to say that historical erasure is not real: it is so pervasive that he is primed to see it even where it isn't. As the wall reflects back at the people trying to read it, the poem's surface reflects back at the reader and away from the vanishing lyric subject, leaving readers with the task of confronting historical elisions that the singular speaker cannot or will not collectively represent.

Dove and Trethewey, like Komunyakaa and Smith, write formally oriented and outwardly quiet poems; Quashie, indeed, identifies them in the tradition of the Black quiet.[47] Related to its quiet effects, their poetry bears the imprint of scientifically backed impersonality—both indirectly through the machinery of the formal poem and directly through appropriating mathematical and scientific concepts to explore poetic knowledge. Dove and Trethewey transform Eliotic scientific impersonality into material for quiet self-exploration. Dove's "Geometry," for instance, follows the steps of a theorem only to insist on the truth of poetic forms of knowledge anchored in the vantage point of her Black woman speaker:

---

[45] Yousef Komunyakaa, *Pleasure Dome: New and Collected Poems* (Middletown, CT: Wesleyan University Press, 2001), 234.
[46] Komunyakaa, *Pleasure Dome*, 235.   [47] Quashie, *The Sovereignty of Quiet*, 107, 132.

> I prove a theorem and the house expands:
> the windows jerk free to hover near the ceiling,
> the ceiling floats away with a sigh.
>
> As the walls clear themselves of everything
> but transparency, the scent of carnations
> leaves with them. I am out in the open
>
> and above the windows have hinged into butterflies,
> sunlight glinting where they've intersected.
> They are going to some point true and unproven.[48]

The poem's stanzas are structured as a three-step theorem, from proposition to development to conclusion. Formally, the poem moves to pin down a proposition, but it begins with "prove" and ends with "unproven," replacing mathematical steps with poetic logic that eludes or negates "proof." In the processes of undoing provability, the poem renders its subject at once intimate and unknowable. "I prove" sets the speaker up as authoritative and solid, a stable point or coordinate in the visualizable space of her house. But the poem plays with the illusion of knowability; the moment we can prove that the house expands, the boundaries of what we know have expanded, ramifying outward into an abstract space of unprovability. As the house's and poem's spaces become abstract and ultimately without coordinates, so does the speaker's interior space. Just when she announces herself as most available to the reader trying to follow her proof—"I am out in the open"—she vanishes in the space between stanzas 2 and 3, never to say "I" again. The poem declares itself as transparent—"the walls clear themselves of everything / but transparency"—yet points to a paradox within the concept of transparency that Komunyakaa's "Facing It" also explores: a fully transparent object is, literally, invisible. Rather than exercising the white male lyric privilege of assumed disembodiment, Dove offers a playful, even joyful, experience of disembodiment in a deliberate act of self-abstraction. In "Geometry," subjectivity becomes a window into understanding the knowledge the poem points us to, which is precisely the value of *not* having full knowledge.

Trethewey's "Theories of Time and Space," a title engaging Einstein, similarly personalizes and abstracts the lyric subject at once. The lyric "I" lives in the depersonalized yet intimate "you" addressee, to whom the speaker ostensibly gives directions, establishing herself as an authority over the space of the

---

[48] Rita Dove, *Selected Poems* (New York: Vintage, 1993), 17.

poem and an authority on loss: "You can get there from here, though / there's no going home." As the poem progresses through its neat, regimented couplets that act as "one mile markers ticking off // another minute," the lyric voice becomes increasingly personal and intimate, the "you" more obviously an unstated "I" who has specific memories of these places. The poem ends with the conceit of a souvenir picture snapped next to a boat in the tourist location of Ship Island: "the photograph—who you were— / will be waiting when you return."[49] Trethewey's poem, like Dove's, plays with the structure of the scientific method; "Follow this to its natural conclusion" is also a proof. As in "Geometry," the speaker's site of authority moves from outward scientific observation to inward contemplation. Precisely when the speaker tells us she is no longer who she was in a photograph, we understand the most about her inner nostalgic state—only to experience the lyric subject as a moving target, morphing across time. This poem, which prefaces Trethewey's elegiac *Native Guard* (2006), is almost a self-elegy for past versions of the speaker. Yet the numerical ambiguity of "you" in English brings in a public, collective possibility that the poem's historical context corroborates. In 2005, Hurricane Katrina, which disproportionately affected the U.S.'s Black population, destroyed 90 percent of the structures in Gulfport and almost fully submerged the tourism-based Ship Island. The poem laments the interwoven losses of times and places, real and imagined. Both in the abstract logic of cosmic time and space, and also, more concretely and more devastatingly, in the materials that made up home, there can be no return. Through its self-address to an ambiguous "you," the poem addresses this collective context while maintaining privacy and unknowability in the speaker, who is never conflated with either a simply individual or a simply collective position.

Like Dove and Trethewey, Smith's speakers dislocate the unspoken whiteness of scientific authority in part by locating the authority of science in personal experience. Smith's lyric voices, at once authoritative and questioning, repurpose perplexing concepts from astrophysics to describe interiority and relationality. The title poem, "Life on Mars," begins by casually reworking an astronomical theory into a poetic metaphor for the ways intimacy and illegibility dually structure human relations: "Tina says what if dark matter is the space between people / When what holds them together isn't exactly love, and I think / That sounds right" (*LOM*, 37). Smith uses dark matter to explain, in affective terms, how interpersonal distance works; dark matter, in brief, holds the "stuff" of the universe together but is not visible. The poem turns

---

[49] Natasha Trethewey, *Native Guard* (Boston, MA: Mariner Books, 2007), 3.

dark matter into an ethical tool, with its mysterious bonds that establish inscrutable space between objects and people in the universe. Smith has spoken of the link between planetary devastation and racial violence in terms of a problem with relations: "We have an impact upon the earth that is analogous to the ways that we treat one another. Consuming, harnessing. These are the things we do to the planet. These are the things that institutions do to individuals."[50] "Life on Mars" uses the dark matter metaphor to explore these "ways that we treat one another." Dark matter explains how "a father…kept his daughter / Locked in a cell for decades"; U.S. soldiers' abuse of Iraqi prisoners; and how U.S.-led neo-imperialism devastates "the earth / We plunder like thieves" as the planet floats in cosmic darkness (*LOM*, 41). The extraterrestrial dimensions of *Life on Mars*, the very scales that would seem to eclipse worldly problems, engage legacies of racism in the U.S. without losing sight of the unfathomable speakers who bring those legacies in and out of focus. Through astronomy, Smith explores nebulous and often violent relations between people without eclipsing the "I" who speaks, imbuing her speakers with opacity to keep them from being conflated with overdetermined political positions. By fusing elegiac lyric, astrophysics, and science fiction—all of which present epistemological impasses—Smith explores public, social events that shape interior experience without offering those personal experiences for representative display.

Life on Mars is publicly pitched, at once a scrutiny of the nation in the twenty-first century and a meditation on cosmic landscapes that contain everyone. In its merging of U.S. politics and space science, the volume implicitly engages with legacies of racism. White supremacist ideology and social structures have, after all, shaped the particular aspirations of space science in the U.S. Lynn Spigel emphasizes the inseparability of space science and white supremacy, specifically as it is enacted through imaginative genres. Spigel argues that in the 1950s–1960s, "whites secured their power through the colonization and control of space" in ways that mirrored control of terrestrial spaces, particularly suburbs (themselves a kind of outlier space). She adds,

> it was not just that whites dominated *physical geographies* through racist zoning laws, transportation policies, and other practices of segregation, they also dominated the culture's *imaginary geographies* of the universe at large.

---

[50] Tracy K. Smith, "'Moving toward What I Don't Know': An Interview with Tracy K. Smith," interview by Claire Schwartz, *Iowa Review* 26:2 (Fall 2016), https://iowareview.org/from-the-issue/volume-46-issue-2-—-fall-2016/moving-toward-what-i-dont-know-interview-tracy-k-smith.

Indeed, in order to maintain and reproduce its power a group must not only occupy physical space, but it must also occupy imaginary space (the space of stories, of images, of fantasy).[51]

Life on Mars reimagines extraterrestrial space as a site to explore inner geographies of Black subjects. It also has an archival impulse to recover and document the life of William Floyd Smith. Many African American contributions to the space program are undocumented, overlooked, or forgotten. Analyzing the Black press's responses to NASA since the 1960s, Spigel notes NASA's tokenizing hiring of a few Black employees, even as the space project still "implicitly endorsed a scientific culture based on segregation, a culture that mimicked the racial division of populations back on earth." Facing criticism for its "less than democratic racial bias" as it tried to present itself as an agency exemplifying the "democratic, free world," NASA responded by hiring African Americans for menial tasks in the agency.[52] Black women were particularly underrepresented. In the 1990s, as space became a canvas to imagine white American ideals of a colorblind post-racial utopia, NASA extensively photographed Dr. Mae Jemison, the first Black woman in space.[53] Where white male astronauts are generally photographed in groups, Jemison is almost always pictured alone or, occasionally, in the company of one white woman astronaut. Jemison's years in NASA correspond to the launch of Hubble. NASA tokenized Jemison from 1987 until her retirement from the agency in 1993; Hubble was launched in 1990, and its mirror corrected in 1993. In putting Hubble at the center of a volume commemorating a Black engineer who worked for NASA in the early 1990s, *Life on Mars* engages with the politics and aesthetics of the visual, ranging from the photographic display of a Black woman astronaut to the sublime aesthetics of Hubble photography that, as Kessler notes, visually recall nineteenth-century depictions of the "Wild West."[54]

---

[51] Lynn Spigel, *Welcome to the Dream House: Popular Media and the Postwar Suburbs* (Durham, NC: Duke University Press, 2001), 145.

[52] Spigel, *Welcome to the Dream House*, 148. Spigel adds, "Over the years NASA continued to respond to racism charges by pointing to its integrated workforce and eventually by appointing one black man, Maj. Robert Lawrence Jr., to the astronaut program. Nevertheless, Over the course of the 1960s, African Americans protested not only the white bias of the astronaut programs, but also they claimed that NASA and its contracted industries tended to employ backs in menial jobs rather than as high-ranking professionals. As we shall see, the corporate culture around NASA was also segregationist, making it difficult for black aerospace workers (of any rank) to live in the areas where whites lived or to participate fully in benefits bestowed by the space program" (148).

[53] Dr. Jemison joined NASA in 1987. She spent eight days in orbit on the STS-47 mission from 12 to 20 Sept. 1992.

[54] Kessler argues that Hubble images are surprisingly similar to nineteenth-century works of art that "portrayed the awe-inspiring and unfamiliar western scenery in the visual language of the sublime" (*Picturing the Cosmos*, 3).

In "The Universe Is a House Party," Smith offsets Wild West expansionist rhetoric with lyric intimacy. In doing so, she challenges the way in which the U.S. universalizes itself by backing expansion with scientific rhetoric. The poem opens with a hypothesis, offering the leftover materials of a house party as poetic evidence of a proposition about entropy:

> The universe is expanding. Look: postcards
> And panties, bottles with lipstick on the rim,
>
> Orphan socks and napkins dried into knots.
> Quickly, wordlessly, all of it whisked into file
>
> With radio waves from a generation ago,
> Drifting to the edge of what doesn't end,
>
> Like the air inside a balloon.
>
> <div align="right">(<i>LOM</i>, 13)</div>

Smith brings a commonplace simile that astronomers often use to make their work legible—the universe is expanding "like...a balloon"—to remind us of what we don't know, the poetic and paradoxical image the simile illuminates: "the edge of what doesn't end." Poets and scientists often use metaphor in inverse of one another: scientists to render difficult concepts legible and poets to render ordinary concepts tantalizingly *il*legible. Through her work with scientific similes, Smith recasts everyday relations—among objects, between the invited and the uninvited, and between eras—with the strange sheen of astronomical discourse. The "radio waves from a generation ago" evoke both the time it takes for the sound of cosmic backdrop radiation to travel in outer space *and* the nostalgic longing in the present for a Cold War era remembered for its belief in the utopian promise of space exploration. The end of the poem critiques the endurance of the Cold War fantasy of American futurity secured by cosmic domination: "Of course, it's ours. If it's anyone's, it's ours" (13). The first-person plural of the poem at once censures manifest destiny in space and elegizes a lost utopian belief in astrofuturism, relegating the belief in an optimistic future of united "humanity" to a past era. The science-fiction trope of aliens as invading racial others reminds us that assimilative versions of a common humanity come about through the dehumanization of others:

> We grind lenses to an impossible strength,
> Point them toward the future, and dream of beings
> We'll welcome with indefatigable hospitality:

> How marvelous you've come! We won't flinch
> At the pinprick mouths, the nubbin limbs. We'll rise,
>
> Gracile, robust. Mi casa es su casa. Never more sincere.
> Seeing us, they'll know exactly what we mean.
>
> (13)

The "we" here are not "citizens of the planet Earth" but, rather, those who assume themselves to be the species norm, welcoming immigrants. The "we" is double-edged, ironic, dystopian, a collective that comes together through its shared self-congratulating performance of polite indifference to otherness. The voice at once speaks as this "we" but knows the alienated condition of the "they," who will, after all, "know exactly" the xenophobia and racism behind the "sincere" and gracious hospitality of the "we" who assume the entire universe is "ours." The alien bodies also convey the ludicrousness of post-racial futurity, distilled in 1960s science fiction like *Star Trek* that imagines humans (who all speak in American accents) will one day conquer the differences that divide "us." The poem channels a species "we" only to reveal it to be a narrowly American collective, evincing the practical and ethical difficulties of speaking as a collective human unit. The poem also suggests the regressive nature of post-racial futurism; the telescope lenses we point into space in ways that feel futuristic ironically show us the deep cosmic past. *Life on Mars* is full of dystopian elements that evoke an old vision of the future that no longer seems possible or desirable. As Smith's version of Charlton Heston (a frequent star of Westerns) puts it, "*That was the future once.../ Before the world went upside down*" (9). Or, as David Bowie has it, "The future isn't what it used to be" (20). The cultural icons Smith quotes, themselves relics of earlier periods, lament now-retrograde futuristic visions from the American past. It is in this blending of elegy and science fiction that the volume approaches its fullest realization of lyric opacity.

5.3 Astrophysics, Science Fiction, and Elegy

Smith's use of the Hubble Space Telescope is, above all, relational. It at once evokes species-wide elegy—Hubble allows us to observe the collective cosmic past—and an individual one, as Smith's elegized father was one of Hubble's optical engineers. The astronomical backdrop enhances the effects of elegy in ways that intensify the opacity of lyric relations. In astronomy, opacity relates to extinction, the phenomenon of electrical radiation diminishing over a given distance as it is scattered and absorbed by nebulous mediums between the astronomical object and the observer. "Why do we insist / He has vanished, that death ran off with our / Everything worth having?" Smith asks of her father, reminding the poem's addressee (her sister Jean) that their inability to see their father in earthbound time and space does not mean that he has literally "vanished" into a cosmic void. Instead, he may simply be beyond the realm of optical comprehension: "He is only gone so far as we can tell. Though / When I try, I see the white cloud of his hair / In the distance like an eternity" (*LOM*, 34). Her father remains just out of the reach of perception and description—a notion accentuated through the astronomical phenomenon that gave us the word "nebulous." "Maybe the dead know" something hazy that the volume doesn't know how to name, "their eyes widening at last, / Seeing the high beams of a million galaxies flick on / At twilight" (10). Even if the dead know something the living cannot, the communication circuit between the "I" and the "thou" breaks down in elegy. The inability of the dead and the living to communicate accentuates the endemic uncertainty in lyric address, deepening an effect Smith explores in lyric at large.

Smith's unlikely blending of genres like elegy and science fiction quietly defamiliarizes the lyric, prompting readers to think about its histories and possibilities anew. Science fiction is a genre thrice removed from Smith in gender, race, and genre. Both astronomical science and science fiction have been overwhelmingly white and male endeavors in film and fiction—or, at least, they are remembered as such.[55] Smith infuses science fiction with elegy to personalize a mode that is almost never associated with interiority. The loss of William Floyd Smith reverberates in the presence of a genre that he loved. The volume takes its title from Bowie's 1971 song "Life on Mars?," which describes a girl underwhelmed by the cinema playing the same films over and over and wonders if she might find life on Mars, as there is nothing remarkable on Earth.[56] Bowie, self-fashioned as an otherworldly being, makes cameos throughout the volume. As the only figure in *Life on Mars* who will "never

---

[55] In a roundtable on Afrofuturism, Sheree Renée Thomas corrects Mark Dery's point that "few African Americans write science fiction." She replies, "[W]hen people ask questions about the absence of black sci-if writers, the answer usually is, 'They exist; you just don't know them.'" She lists Jewelle Gomez, Wanda Coleman, Ishmael Reed, Amiri Baraka, Henry Dumas, Charles R. Saunders, Steven Barnes, Mary C. Aldridge, and others "who would have been very surprised to learn that they weren't writing science fiction." See Tiffany E. Barber, "25 Years of Afrofuturism and Black Speculative Thought: Roundtable with Tiffany E. Barber, Reynaldo Anderson, Mark Derry, and Sheree Renée Thomas," *TOPIA: Canadian Journal of Cultural Studies* 39 (2018): 136–44, here 144.

[56] David Bowie, "Life on Mars?", recorded June–July 1971, track 4 on *Hunky Dory*, RCA Records, vinyl LP.

die," Bowie watches over the volume with cosmic distance and benevolence, his unfathomability modeling a form of opacity. Bowie becomes a way of understanding rather than an object to understand; he is a figure "aching to make us see" even as the poems never let us fully see him:

> After dark, stars glisten like ice, and the distance they span
> Hides something elemental. Not God, exactly. More like
> Some thin-hipped glittering Bowie-being—a Starman
> Or cosmic ace hovering, swaying, aching to make us see.
>
> (*LOM*, 19)

The passage riffs on similes, beginning with the deliberately clichéd "stars glisten like ice," then teasing its own obvious simile through the zany claim that interstellar space hides not God but rather something "more like" (and still not exactly) Bowie. The continual simile revisions keep the scene out of focus for the reader, whose hermeneutical strategies (and search for poetry's cosmic truths) are thwarted. The verb "see" in the passage has no object: the phrase "aching to...see," as opposed to bringing objects into resolution, describes a process rather than an end goal, a way of reading rather than a quest to unveil objects and people that perhaps cannot or should not be fully on display. The passage also suggests that Smith's father has become a sort of eternally cosmic figure, Bowie's indiscernible but present "Starman waiting in the sky." Smith demonstrates how contact with astrophysics and science fiction transforms elegy's impossible propositions—to reorganize time and reanimate the dead—into otherworldly aspirations.

The temporal knots and gaps in knowledge in both science fiction and elegy are Afrofuturist moves. In *Ordinary Light*, Smith echoes the relationship of the often-illegible African American past to future vistas; she recalls how her father, a patriot and a Reaganite, seldom spoke of his childhood in the violent and segregated state of Alabama. Smith asks, "Why was it so much easier to call out to the future than the past? I still couldn't bring myself to actually talk with my parents about what it must have meant to grow up in an age of racial violence" (*OL*, 73). Much of the volume portrays Smith's process of searching for knowledge of his past in the Jim Crow South, which is almost as opaque to his daughter as his present place in the cosmic afterlife. The illegibility of the African diasporic past poses a problem for Afrofuturist writers; as Dery argues, Afrofuturism "gives rise to a troubling antinomy: Can a community whose past has been deliberately rubbed out, and whose energies have subsequently been consumed by the search for legible traces of its history, imagine possible futures?"[57] The charged relationship between the past and a hard-to-imagine future itself speaks to the difficulty of modern elegy. Commenting on what I am suggesting is science fiction's cousin genre, Jahan Ramazani argues that modern elegy struggles to make sense of the relationship between past and future, complicating or even disposing of traditional elegy's final step of consolingly imagining a future after loss. Modern elegy becomes *anti*-consolatory as elegists seek "a vocabulary for grief in our time—elegies that erupt with all the violence and irresolution, all the guilt and ambivalence of modern mourning."[58] These "anti-elegies" make possible futures harder to imagine, if not obsolete. Smith brings science fiction into close contact with elegy to heighten the otherworldly effects of lyric opacity. Seo-Young Chu offers the only study to date on the intersections of these genres. In *Do Metaphors Dream of Literal Sleep?*, she contends that science fictional and "[l]yric voices speak from beyond ordinary time"[59]—an effect that is particularly dramatic in elegy. Traditional elegy pays homage to the dead by reflecting on their lives from an emotionally unsettled present moment, ultimately to imagine a future enhanced by remembrance. The epistemological impasses of science fiction and elegy help Smith to contend with parts of the past that cannot be known—even as the cosmic setting offers new possibilities for modern elegy, in which futurity is itself an obsolete goal. Smith's cosmic elegies are not, after all, future-oriented; they are instead relational, drawing connections not just between people and their spatial environments but also between people and their temporal ecosystems.

"Sci-Fi" opens *Life on Mars* with a fusion of science fiction and elegy that demonstrates opaque relations across different times: "There will be no edges, but curves. / Clean lines pointing only forward," the poem metapoetically announces in clean lines in sharp couplet stanzas (*LOM*, 7). In an event at Stanford University combining poetry and astrophysics, Smith explained that the poem, like science fiction in general, is an elegy for a defunct future. Rewatching these films as an adult led Smith to realize that "the future" these films depicted "is not so much futuristic as it is an homage and an elegy to a present that the filmmaker was in the process of living.... This is no longer what the future looks like." She thinks specifically of Stanley Kubrick's *2001: A Space Odyssey*, a film she watched repeatedly with her father as a child, and its

[57] Dery, "Black to the Future," 180.
[58] Jahan Ramazani, *Poetry of Mourning: The Modern Elegy from Hardy to Heaney* (Chicago, IL: University of Chicago Press, 1994), ix.
[59] Seo-Young Chu, *Do Metaphors Dream of Literal Sleep?* (Cambridge, MA: Harvard University Press, 2010), 13.

now-outmoded vision of a linear future, when "now our experience of space is more curved."⁶⁰ The poem is full of sharp, linear designs from this past represented in its neat couplet structure and dystopian progress narrative. The species "we" in the poem inhabits a retrograde vision of futurity, a world where gender and sex will be obsolete:

> And yes, we'll live to be much older, thanks
> To popular consensus. Weightless, unhinged,
>
> Eons from even our own moon, we'll drift
> In the haze of space, which will be, once
> And for all, scrutable and safe.
>
> (*LOM*, 7)

The tone here is remarkably complex, conveying nostalgia for a now-defunct vision of a future from a past era that believed in progress. In fact, Erin Ranft suggests that "Sci-Fi" is an Afrofuturist work that "imagines a space and time where harmony may exist amongst all peoples."⁶¹ Yet the final lines have a dystopian quality. The last two lines of the poem break in the middle of the words "once / And for all," severing a phrase that echoes the unfulfilled promise at the end of the Pledge of Allegiance to provide "justice for all." The end of the poem points to injustices in the present world that does not offer safety "for all."

The volume directly confronts racist violence in its longest poem, "They May Love All That He Has Chosen and Hate All That He Has Rejected," which emphasizes the generic relationships between science fiction and elegy. Smith reflects that the first poem in which she broke her "silence" on the trauma of racism was in this poem. Its central section, "In Which the Dead Send Postcards to Their Assailants from America's Most Celebrated Landmarks," is a cluster of persona poems in which dead victims of hate crimes address their killers in disembodied voices from beyond the world. These voices are the stuff of science fiction, the only voices that resound from the volume's opaque cosmic "it." These persona poems drive a clear wedge between author and lyric speaker; in them, there is no mistaking the "I" as autobiographical. In the opening prose poem, nine-year-old Brisenia Flores, murdered in her family home by the white supremacist Minutemen American Defense group in 2009, addresses one of her killers familiarly from the Statue of Liberty: "Dear

---

[60] Smith, "Imagining the Universe."
[61] Erin Ranft, "The Afrofuturist Poetry of Tracie Morris and Tracy K. Smith," *Journal of Ethnic American Literature* 4 (2014): 75–85, here 80.

Shawna, How are you? Today we took a boat out to an island. It was cold even though the sun was hot on my skin" (*LOM*, 49). The word "skin," as well as the visual medium of the postcard, emphasizes how deeply these poems are about racial violence. These ekphrastic poems use the politics and aesthetics of the visual to explore strange and sometimes violent relations between people, as well as between people and their national and planetary environments.

"The Museum of Obsolescence" imagines another dystopian future that comments on the dystopian present. The poem is set in a museum full of forgotten and useless artifacts from the human species' time on Earth: "There's green money, and oil in drums. // Pots of honey pilfered from a tomb." The same capitalist system that led to planetary devastation endures in this astrofuturist vision; outside of the Museum of Obsolescence, "vendors hawk t-shirts, three for eight." The final exhibit item before the museum's exit "is an image of the old planet taken from space," at once a poignant relic of a bygone era and a piece of mass-produced junk (14). The "image" is one of the first photographs of the whole Earth viewed from space, prominently displayed in glossy magazines for public consumption. The "image of the old planet" displayed by the exit of the Museum of Obsolescence is not so different from the cheap, mass-produced souvenirs of capitalism that wait in the shop outside. Smith demonstrates how that planet is now an elegiac artifact in its own right, recalling a bygone age of space utopianism. To the human (or post-human) visitors of the museum, the image is now unrecognizable; it is just *an* old planet, not *our* old planet. The poem is an anachronistic elegy for Earth written from the present. But above all, it is an elegy for the 1960s and 1970s American generation that *Blue Marble* came out of, a period where some still imagined humans could collectively come together to make the planetary future work for our species. That future feels increasingly a relic of the past.

The reproduced Earth photograph also laments a lost connection to Earth as a home, a sense intensified by the image's otherworldliness. One of *Life on Mars*'s most important intertexts, *2001: A Space Odyssey*, ends with a human fetus in space untethered from planet Earth, at which it gazes. Ange Mlinko finds the theme of a child unmoored in the cosmos throughout *Life on Mars*, attributing it to Smith's sense of herself as newly orphaned.[62] Despite the intimate and personal source material, the poem's lyric "I" is subsumed into an impersonal lyric "we," the object of Smith's grief outsourced to "maps of fizzled stars" and other obsolete paraphernalia that "watches us watch it" with

---

[62] Ange Mlinko, "The Lyric Project: On Tracy K. Smith and Cathy Park Hong," *Parnassus: Poetry in Review* 33:1–2 (2013): 417–26 and 432, here 422.

"[o]ur faulty eyes" (*LOM*, 14). Through its collective scope—its use of "we" rather than "I," the public and ceremonial space of a museum, the dystopian and ironic tone—the poem furtively achieves an opaque space for private grief by amplifying that grief into an experience of losing an entire shared world.

Even Smith's elegiac villanelle "Solstice" is infused with science fiction. The villanelle form has only one possible future: the assimilation of the poem's two repeated lines and rhymes into a final quatrain. The mathematical structure of "Solstice" formally enacts a closed future; we know the nightmarish future before we finish reading. Following Bishop's technique of tangling different scales of loss in interlocked sounds (exemplified in "One Art"), Smith inflates Bishop's geographical interest in scale to planetary proportions. "Solstice" is set during the time of year when Earth is most off-kilter on its axis, a metaphor for American power imbalance on the global stage as well as for tragedies and retributions out of proportion to one another. On 15 January 2009, several weeks after the winter solstice in the northern hemisphere, US Airways Flight 1549 ditched in the Hudson River shortly after it collided with a flock of geese on takeoff from JFK International Airport. The comparatively tiny birds bringing down the huge jetliner resulted in what the poem presents as a disproportionate response: "They're gassing geese outside of JFK," it matter-of-factly begins (*LOM*, 43). From there, the villanelle goes on to confuse events, times, and scales, moving from the geese to a larger horror in New York City's skyline: "They're going to make the opposition pay. / (If you're sympathetic, knock on wood.) / The geese were terrorizing JFK." The bizarre conflation of geese with terrorists suggests that the trauma of 9/11 distorted both national memory and the U.S's subsequent military actions in the Middle East. The deranged, associative logic of the villanelle offers no clear path to disentangle discrete events: "They're gassing geese outside of JFK. / Tehran will likely fill up soon with blood. / The *Times* is getting smaller by the day." In the third tercet, the third line mutates into "The *Times* reported 19 dead today," and finally, in the quatrain, Smith delivers the inevitable and bleak conclusion: "Our time is brief. We dwindle by the day." In its closed, predictable future, the villanelle is a remarkably transparent form. Like Bishop, Smith plays the villanelle's transparent course against the opacity of a speaker who minimizes her reactions to the outrageous dilations of scale that she records. As "One Art" tumbles out of the speaker's control, leaving her devastation exposed at the end, Smith's lyric "we" begins with factual observations but ends with an admission of our collective doom. The lyric "I" channeled into a lyric "we" suggests, too, that all the people and species of the poem, however unintelligible to one another, share a fate on a planet in distress, in which violence to the planet comes out of the same systems that underlie the violence of American institutions toward their own marginalized citizens and people around the globe. Smith's collectively pitched elegies like "The Museum of Obsolescence," "Sci-Fi," and "Solstice" suggest that the only possible futures are either nonexistent or undesirable. They are revealed to be lost pasts, or post-racial dystopias, or foregone conclusions of planetary doom. Yet even the logic of the villanelle suggests that futurity is the wrong question for elegy. The villanelle blends times, never coming to terms with its own pasts, tangling scales of and moments in time in a kind of dream (or perhaps nightmare) logic of temporal relations.

Science fiction and elegy interanimate one another's temporal confusions in ways that render the past and the future equally unknowable, intensifying the mysteriousness of lyric subjects who become difficult to place not only in space but also in time. In "My God, It's Full of Stars," Smith's lyric "I" imagines going back to a time before she existed in an attempt to apprehend her father. She personalizes a seemingly universal concept, "time,"

> Which should curl in on itself and loop around like smoke.
> So that I might be sitting now beside my father
> As he raises a lit match to the bowl of his pipe
> For the first time in the winter of 1959.
>
> (10)

The smoke is a time-traveling device that loops the poem back to an impossible setting, one that inverts the subject and object of elegy: rather than commemorating a dead parent, the "I" travels back to a time when he was living before she existed to watch him, undetected herself, in an everyday moment. Another section of the poem describes William Floyd Smith in 1950s Sunflower, Alabama, before his daughter's birth: "Sometimes, what I see is a library in a rural community. / All the tall shelves in the big open room. And the pencils / In a cup at Circulation, gnawed on by the entire population" (8). The tone is warm, inviting, and straightforward; an "I" reports an everyday scene to the reader with great clarity, capturing even the zoomed-in, hyper-focused visual detail of teeth marks on pencils. But the "I" leaves out the broader historical context that William Floyd Smith also elided in family conversations—for example, the Jim Crow laws that would have prevented him from using the circulation services of Washington County, Alabama's public libraries.[63] The intense, precise visual details belie all that is left unsaid across generations.

No poem more poignantly explores the difficulty of lyric relations than Smith's extended elegy for her father, "The Speed of Belief," protectively folded into the center of *Life on Mars*. Much of the elegy is spent trying to get him to speak. Against the backdrop of an apparent cosmic void, she brings her father in and out of focus by fluctuating between direct and indirect address. This vacillation plays with presence and absence, tracking the fits and starts of the temporality of grief by modulating the perceived distance of the dead. "But where does all he knew—and all he must now know—walk?" Smith asks before training such questions directly on her father (*LOM*, 29). Apostrophe, literally a "turning away," gives the effect of a speaker turning her back on the reader to face an absent listener. Smith tempts her absent listener with obscure questions that he, as a scientist, a person of faith, and now a kind of ghost, is uniquely situated to answer: "And will it drag you back / As flesh, voice scent?" (33). The poetic question adds urgency to the already commanding structure of apostrophe, which always at least implies an imperative; the apostrophic question is simultaneously a request for a reply and a disappointment, highlighting the impossibility of a future situation in which the dead might respond.

But Smith's cosmic elegies offer some consolation by making futurity the wrong issue. The elegiac cover image of *Life on Mars* suggests an opaque cosmic whole held together not only in space but also in time. The Hubble image of the Cone Nebula displays an interstellar cloud of gas and dust jutting through stars of various magnitudes and distances. The nebula is opaque, absorbing most visible light along its black trunk and scattering some in a marbled smear of pink and green at its crown. This partial opacity amplifies the image's mysteriousness, frustrating our ability to see inside of the cloud where molecules are—or were—fusing to form new stars. The Cone Nebula blurs past, present, and future: twenty-first-century observers of the stellar nursery see evidence of the births of stars that are now aged or dead. This temporal scrambling suggests the time warp in which Smith found herself while working on the poems. When her father fell ill and died during Smith's composition of the volume, Smith was also pregnant with her daughter, giving her a sense, as she described in a 2011 PBS interview, that "not only is

---

[63] While most libraries in 1950s Alabama were segregated prior to sit-ins in the 1960s, some Alabama libraries allowed limited services to Black patrons, such as use of the reading room but not circulation services.

there this ever-after that our loved ones disappear into, but there's some source that might be generating other people, other...loves."[64] Smith's time-traveling elegies trace the strange relational force of love across violent eras in both human time and cosmic deep time. They model at once the ethical possibilities and the challenges of lyric opacity, laying bare the dead's frustrating, heartbreaking refusal—or inability—to make themselves legible to life on Earth.

[64] Tracy K. Smith, "'Life on Mars' Author Explores Humans' Relationship with Universe Through Poetry," interview by Gwen Ifill, PBS Newshour, 16 May 2011, http://www.pbs.org/newshour/bb/entertainment-jan-june11-tracysmith_05-16/.

Coda

Poems in Space

Above all, the extraterrestrial offers another perspective on Earth. Each chapter of this book considered at least one lyric poem that describes planet Earth photographed in space between 1966 and 1972: the NASA-commissioned celebratory poems by Archibald MacLeish and James Dickey in Chapter 1; James Merrill's elegy for Elizabeth Bishop as well as her own "In the Waiting Room" in Chapter 2; Agha Shahid Ali's formative "Lunarscape" in Chapter 3; Seamus Heaney's "Alphabets" in Chapter 4; and Tracy K. Smith's "The Museum of Obsolescence" in Chapter 5. The image is an enduring reminder of the enmeshed scales of personal, national, global, and planetary experience. It is also, I think, an elegy.

In 1990, astronomer Carl Sagan updated this iconic lyric shot from a more distant and alien vantage point, one that specifically invokes the afterlife. He convinced NASA to turn the Voyager 1 spacecraft around to photograph the "pale blue dot" of Earth from Saturn, a distance of 6 billion kilometers. The gesture was in part poetic, as Sagan, a poetry fan, recognized: the sphere of Saturn is the farthest point in Dante's journey through the Ptolemaic cosmos as he looks back through seven spheres at the distant globe of Earth. In Canto XXII of the *Paradiso*, Beatrice instructs Dante to "Look downward, and contemplate, what a world / Already stretch'd under our feet there lies."[1] Dante responds with the gaze back at Earth from the heavens that characterizes so much of postwar space exploration:

> I straight obey'd; and with mine ye return'd
> Through all the seven spheres; and saw this globe
> So pitiful of semblance, that perforce
> It moved my smiles.[2]

---

[1] Dante Alighieri, *Paradiso* XXII, 124–5, trans. Henry Francis Cary (Roseville, CA: Dry Bones Press, 2000).

[2] Dante, *Paradiso* XXII, 129–32.

"Pale Blue Dot" weds Dante's katabasis to the messianic American story of saving the world's soul. The photograph went on to inspire its own poem, Maya Angelou's "A Brave and Startling Truth," which evokes the *Pax Americana* through echoes of the Declaration of Independence:

> We, this people, on a small and lonely planet
> Traveling through casual space
> Past aloof stars, across the way of indifferent suns
> To a destination where all signs tell us
> It is possible and imperative that we learn
> A brave and startling truth...[3]

Written for the United Nations, the poem was blasted into Earth orbit aboard NASA's Orion in 2014 following Angelou's death, in an elegiac tribute to Angelou. Her great-grandchildren gifted the broadside to Charles Bolden, the first Black NASA administrator. This event was, in part, a tokenizing event to help NASA convey the American dream and its help in realizing the dawning of a post-racial utopia. But it is also a welcome change to see not a godlike white Apollo figure mastering Earth from above but, rather, a Black woman's face in profile above the planet, looking not down at Earth but, rather, out into space at what we cannot see or know outside the frame, in death.[4]

Angelou's poem is one recent example of what Leonard describes as "the strange practice of sending poems into space," into the very realm of gods.[5] The first of these poems, Thomas Bergin's "For a Space Prober," his utopian poem for world peace, was launched in 1961 to orbit for 8,000 years. Since then, NASA has sent thousands of poems into Earth orbit and to other planets, including the 1,100 haiku currently orbiting Mars. The lyric has even traveled imaginatively to Pluto; in July 2015, NPR's website featured a video captioned "Pluto Mission Gets a Poetic Tribute." As the video rolls through the New Horizon probe's first topographical photographs of Pluto, Ray Bradbury's

---

[3] Maya Angelou, "A Brave and Startling Truth," in *Celebrations: Rituals of Peace and Prayer* (New York: Random House, 2006), 17.

[4] Sarah Loff, "Poem by American Matriarch Flown on Orion Presented to NASA Administrator," *NASA*, 6 Apr. 2015, https://www.nasa.gov/content/poem-by-american-matriarch-flown-on-orion-presented-to-nasa-administrator. The poem was also read at the annual Verse in Universe event: Maria Popova, "A Brave and Startling Truth: Astrophysicist Janna Levin Reads Maya Angelou's Stunning Humanist Poem That Flew to Space, Inspired by Carl Sagan," *BrainPickings* (9 May 2018), https://www.brainpickings.org/2018/05/09/a-brave-and-startling-truth-maya-angelou/.

[5] Philip Leonard, "Message to the gods: the space poetry that transcends human rivalries," *The Conversation*, 15 Nov. 2017, https://theconversation.com/message-to-the-gods-the-space-poetry-that-transcends-human-rivalries-86572.

1970s recording of his poem "If Only We Had Taller Been" plays in the background—perhaps an appropriate choice for the recently declassified planet Pluto, and a planetary elegy for a poem named for the god of the afterlife. Most recently, National Poet Youth Laureate Amanda Gorman's and Poet Laureate Joy Harjo's words have traveled on the Lucy Asteroid Probe's Time Capsule Plaque that advertises prophetic "messages to future humans."[6] Both Gorman and Harjo have evoked the whole Earth seen from space. Gorman's poem "Our Purpose in Poetry: Or, Earthrise" suggests that poetry and *Earthrise* do the same work, allowing humankind "to dream a different reality" as we attempt to respond to climate change.[7] Harjo has also invoked a whole Earth photograph, speaking movingly of *Blue Marble* in an episode of the podcast *On Being*. The image, for her, visualizes what she calls the "whole of time," including the lifespan of the Earth, the historical time that includes the Muscogee Nation before and after and European colonization, and the lifespans of individual humans and their ancestral relations. Harjo reflects, "that NASA image of the Earth... that showed the Earth as a beautiful, beautiful being, so powerful... gave us a perspective which going into a larger time and place can... it gave us that glimpse into even another kind of time."[8] This kind of time neighbors the strange temporality of elegy, which blends past, present, and future.

Launching lyric poetry into space is almost a type of space burial, the practice of sending human remains into orbit. *Star Trek* creator Gene Roddenberry's went up twice in the 1990s, and SpaceX has begun offering "funeral flights." When William Shatner lifted off on Jeff Bezos's Blue Origin rocket in 2021, he relayed the experience as a kind of self-elegy. Tanner Stening records the elegiac nature of Shatner's trip in his occasional poem on the event: "What you see is *black*, he said. *Is that death?*" Reminding us that "Captain Kirk had died three times," Stening suggests something of Shatner's near immortality, contrasting the black void of space to the lively blue planet that mesmerized Shatner below.[9] And there is, perhaps, something fundamentally elegiac in the way space exploration fails to make us eternal.

---

[6] See Elizabeth Bowell, "NASA's Lucy asteroid probe has a time capsule plaque with messages to future humans," *Space.com*, 15 Oct. 2021, https://www.space.com/nasa-lucy-asteroids-plaque-for-the-future.

[7] Amanda Gorman, "Our Purpose in Poetry: Or, Earthrise," *Sierra Club: Los Padres Chapter*, 3 Feb. 2021, https://www.sierraclub.org/los-padres/blog/2021/02/earthrise-poem-amanda-gorman.

[8] Joy Harjo, "The Whole of Time," *On Being with Krista Tippet*, 13 May 2021, https://onbeing.org/programs/joy-harjo-the-whole-of-time/.

[9] Tanner Stening, "This Covering of Blue," *Rattle*, 17 Oct. 2021, https://www.rattle.com/this-covering-of-blue-by-tanner-stening/.

"We came in peace for all mankind," the Apollo 11 lunar plaque reads, immortalizing humanity in an American image even while recording in the past tense the species' and empire's impermanence. Apollo 11's commemorative lunar artifacts are now relics of a declining empire. The flag the mission left on the moon is no longer visible to telescopes, having toppled over in the engine blast from Apollo 11's ascent. Assuming it is still there, it has lost its colors and distinctive stars and stripes that wrote its destiny into the cosmos. Against the impermanence of America's cosmic footprint, the lyric, as it is culturally imagined, might seem a way to outlive "the gilded monuments."

For contemporary poets, the lyric rarely offers such assurances of immortality. The poet's task, W. H. Auden reflected, becomes more difficult as cosmological models have stripped away even cosmic permanence. His elegies refuse to offer consolation through appealing to cosmic eternity. "The stars are not wanted now; put out every one, / Pack up the moon and dismantle the sun," Auden wrote in "Funeral Blues," packing away the elegist's cosmic props.[10] Berkeley astronomer Geoff Marcy recently adapted this poem to elegize NASA's Kepler craft, which was tasked with searching for Earthlike planets amid alien stars. When one of its reaction wheels failed, Kepler was retired, prematurely ending its mission to find other worlds that our species might one day inhabit. Marcy wrote:

> Kepler was my North, my South, my East and West,
> My working week, no weekend rest,
> My noon, my midnight, my talks, my song;
> I thought Kepler would last forever: I was wrong.[11]

The parody is meant to be funny, as Auden's "Funeral Blues" almost is, but it also records the serious situation of planetary impermanence. Kepler's malfunction put an end to its distant hope of finding Earthlike planets through which humans could project the species into the future, imaginatively or perhaps eventually literally. Poems in space are meant to outlast the species that made them; they won't save us, but they are poignant self-elegies, cosmic relics of an ancient planet.

---

[10] W. H. Auden, *Collected Poems*, ed. Edward Mendelson (New York: Random House, 2007), 141.
[11] Clara Moskowitz, "Astronomer's Poem Mourns Ailing Kepler Spacecraft," 16 May 2013, *In Brief*, Space.com, https://www.space.com/21179-kepler-telescope-failure-poem.html.

Bibliography

Ackerman, Diane. *The Planets: Cosmic Pastoral*. New York: William Morrow and Co., 1976.
Adams, John. *The Works of John Adams, Second President of the United States*. 10 vols. Edited by Charles Francis Adams. Boston, 1856.
Adorno, Theodor W. "On Lyric Poetry and Society." In *Notes on Literature*. Edited by Rolf Tiedeman, translated by Shierry Weber Nicholsen. New York: Columbia University Press, 1991.
Ahearn, Barry. *Pound, Frost, Moore, and Poetic Precision: Science in Modernist American Poetry*. New York: Palgrave Macmillan, 2020.
Albrecht, Glenn. "Solastalgia: A New Concept in Health and Identity." *PAN: Philosophy, Activism, Nature* 3 (2005): 41–59.
Aldrin, Buzz. *Magnificent Desolation: The Long Journey Home from the Moon*. New York: Three Rivers Press, 2010.
Ali, Agha Shahid. *In Memory of Begum Akhtar*. Kolkata: Writers Workshop, 1979.
Ali, Agha Shahid. "Postcard from Kashmir." Manuscript and typescript drafts, 1979. Agha Shahid Ali Papers. Hamilton College Library Special Collections.
Ali, Agha Shahid. *Ravishing DisUnities: Real Ghazals in English*. Hanover, NH: Wesleyan University Press, 2000.
Ali, Agha Shahid. "A Darkly Defense of Dead White Males." In *Poet's Work, Poet's Play: Essays on the Practice and the Art*. Edited by Daniel Tobin and Pinone Triplett. Ann Arbor, MI: University of Michigan Press, 2008, 144–60.
Ali, Agha Shahid. *The Veiled Suite*. New York: Penguin Books, 2010.
Ali, Agha Shahid. "Agha Shahid Ali: The Lost Interview." Interviewed by Stacey Chase. *Café Review* 22 (Spring 2011).
Ali, Agha Shahid, and Faiz Ahmad Faiz. *The Rebel's Silhouette: Selected Poems*. Amherst, MA: University of Massachusetts Press, 1995.
Alighieri, Dante. *The Divine Comedy of Dante Alighieri*. Translated by Henry Francis Cary. Roseville, CA: Dry Bones Press, 2000.
Allen, Chadwick. *Trans-indigenous: Methodologies for Global Native Literary Studies*. Minneapolis, MN: University of Minnesota Press, 2012.
Anderson, Benedict. *Imagined Communities: Reflections on the Origin and Spread of Nationalism*. London: Verso, 2016.
Angelou, Maya. *Celebrations: Rituals of Peace and Prayer*. New York: Random House, 2006.
Appiah, Kwame Anthony. *The Ethics of Identity*. Princeton, NJ: Princeton University Press, 2005.
Appiah, Kwame Anthony. *Cosmopolitanism: Ethics in a World of Strangers*. New York: Norton, 2010.
Ardent, Hannah. "Imperialism, Nationalism, Chauvinism." *Review of Politics* 7:4 (Oct. 1945): 441–63.
Ardent, Hannah. *Between Past and Future: Eight Exercises in Political Thought*. New York: Penguin, 1977.
Ardent, Hannah. *The Human Condition*. 2nd ed. Chicago, IL: University of Chicago Press, 2013.

Asghar, Fatimah. "Pluto Shits on the Universe." *Poetry* (April 2015). https://www.poetryfoundation.org/poetrymagazine/poems/58056/pluto-shits-on-the-universe.

Associated Press. "50 years ago the Apollo 8 moonshot gave humans a new perspective of Earth." *Washington Post* online (21 Dec. 2018). https://www.washingtonpost.com/lifestyle/kidspost/50-years-ago-the-apollo-8-moonshot-gave-humans-a-new-perspective-of-earth/2018/12/21/998c8db4-02f5-11e9-9122-82e98f91ee6f_story.html.

Auden, W. H. *Collected Longer Poems of W. H. Auden*. New York: Random House, 1965.

Auden, W. H. *Dyer's Hand and Other Essays*. New York: Vintage, 1989.

Auden, W. H. *Collected Poems*. Edited by Edward Mendelson. New York: Random House, 2007.

Axelrod, Steven Gould. "Elizabeth Bishop and Containment Policy." *American Literature* 75:4 (2003): 843–67.

Baer, William. *Conversations with Derek Walcott*. Jackson: University Press of Mississippi, 1996.

Bakhtin, M.M. *The Dialogic Imagination: Four Essays*. Austin: University of Texas Press, 2017.

Banerjee, Rita. "Between Postindependence and the Cold War: Agha Shahid Ali's Publications with the Calcutta Writers Workshop." In *Mad Heart Be Brave: Essays on the Poetry of Agha Shahid Ali*. Edited by Kazim Ali. Ann Arbor: University of Michigan Press, 2017, 20–32.

Baraka, Amiri. *Black Magic: Sabotage, Target Study, Black Art: Collected Poetry, 1961–1967*. Indianapolis, IN: Bobbs-Merrill, 1969.

Baraka, Amiri. "A Post-Racial Anthology?" In *Review of Angles of Ascent: A Norton Anthology of Contemporary African American Poetry*. Edited by Charles Henry Rowell. New York: W.W. Norton, 2013. *Poetry* (May 2013). https://www.poetryfoundation.org/poetrymagazine/articles/69990/a-post-racial-anthology.

Barbauld, Anna Laetitia. *The Poems of Anna Letitia Barbauld*. Athens: University of Georgia Press, 1994.

Barber, Tiffany E. "25 Years of Afrofuturism and Black Speculative Thought: Roundtable with Tiffany E. Barber, Reynaldo Anderson, Mark Dery, and Sheree Renée Thomas." *TOPIA: Canadian Journal of Cultural Studies* 39 (2018): 136–44.

Battersby, Christine. *The Sublime, Terror, and Human Difference*. London: Routledge, 2007.

Bennett, Eric. *Workshops of Empire: Stegner, Engle, and American Creative Writing During the Cold War*. Iowa City: University of Iowa Press, 2015.

Bercovitch, Sacvan. *The American Jeremiad*. Madison: University of Wisconsin Press, 1978.

Bhabha, Homi. *The Location of Culture*. London: Routledge, 1994.

Bishop, Elizabeth. *One Art: The Selected Letters*. Edited by Robert Giroux. London: Pimlico, 1994.

Bishop, Elizabeth. *Geography III*. New York: Farrar, Straus and Giroux, 2008.

Bishop, Elizabeth. *Poems, Prose, and Letters*. Edited by Robert Giroux and Lloyd Schwartz. New York: Library of America, 2008.

Bishop, Elizabeth. *Words in Air: The Complete Correspondence between Elizabeth Bishop and Robert Lowell*. Edited by Thomas Travisano with Saskia Hamilton. New York: Farrar, Straus and Giroux, 2008.

Bishop, Elizabeth. *Poems*. New York: Farrar, Straus and Giroux, 2011.

Blake, William. *The Complete Poetry and Prose of William Blake*. Edited by David V. Erdman. New York: Anchor Books, 1988.

Bowie, David. "Life on Mars." Recorded June–July 1971. Track 4 on Hunky Dory. RCA Records, vinyl LP.

Bowell, Elizabeth. "NASA's Lucy asteroid probe has a time capsule plaque with messages to future humans." *Space.com* (15 Oct. 2021). https://www.space.com/nasa-lucy-asteroids-plaque-for-the-future.
Boyde, Patrick. *Dante, Philomythes and Philosopher: Man in the Cosmos*. Cambridge: Cambridge University Press, 1981.
Boym, Svetlana. *The Future of Nostalgia*. New York: Basic Books, 2011.
Breslin, Paul. *Nobody's Nation: Reading Derek Walcott*. Chicago: University of Chicago Press, 2001.
Brodsky, Joseph. *Less Than One: Selected Essays*. New York: Farrar, Straus and Giroux, 1986.
Brodsky, Joseph. *On Grief and Reason: essays*. New York: Farrar, Straus and Giroux, 1995.
Brodsky, Joseph. *Collected Poems in English*. New York: Farrar, Straus and Giroux, 2002.
Brodsky, Joseph, and Seamus Heaney. "Poetry and Politics: A conversation between Seamus Heaney and Joseph Brodsky." Edited by Fintan O'Toole. *Magill* (31 Oct. 1985). https://magill.ie/archive/poetry-and-politics-conversation-between-seamus-heany-and-joseph-brodsky.
Brodsky, Joseph, Seamus Heaney, and Derek Walcott. *Homage to Robert Frost*. New York: Farrar, Straus and Giroux, 1996.
Brooks, Cleanth. *The Well Wrought Urn: Studies in the Structure of Poetry*. New York: Harcourt Brace, 1947.
Brothers, Dometa Wiegand. *The Romantic Imagination and Astronomy: On All Sides Infinity*. New York: Palgrave, 2015.
Burke, Edmund. *A Philosophical Enquiry into the Origin of Our Ideas of the Sublime and Beautiful*. Edited by Adam Phillips. Oxford: Oxford University Press, 1998.
Burnell, Jocelyn Bell. "Astronomy and Poetry." In *Contemporary Poetry and Contemporary Science*. Edited by Robert Crawford. Oxford: Oxford University Press, 2006, 125–40.
Burt, Stephanie. *The Poem is You: 60 Contemporary American Poems and How to Read Them*. Cambridge, MA: The Belknap Press of Harvard University Press, 2016.
Burt, Stephanie. "Agha Shahid Ali, World Literature, and the Representation of Kashmir." In *Mad Heart Be Brave: Essays on the Poetry of Agha Shahid Ali*. Edited by Kazim Ali. Ann Arbor: University of Michigan Press, 2017, 104–17.
Burt, Stephanie. "Space Travels." *The Nation* (1 June 2017). https://www.thenation.com/article/archive/Czesław-Miłosz%20s-space-travels/.
Byrd, Jodi A. *The Transit of Empire: Indigenous Critiques of Colonialism*. Minneapolis: University of Minnesota Press, 2011.
Byrnes, Mark E. *Politics and Space: Image Making by NASA*. Westport, CT: Praeger Publishers, 1994.
Cameron, Sharon. *Lyric Time*. Baltimore, MD: Johns Hopkins University Press, 1979.
Cantor, Paul A. "Shakespeare in the Original Klingon: Star Trek and the End of History." *Perspectives on Political Science* 29:3 (Apr. 2010): 158–66.
Carrington, André M. *Speculative Blackness: The Future of Race in Science Fiction*. Minneapolis: University of Minnesota Press, 2016.
Carson, Ciaran. "Escaped from the Massacre?" *The Honest Ulsterman* 50 (Winter 1975): 183–6.
Casper, Rob. "Space, Time, and the Poet Sagan." *From the Catbird Seat: Poetry & Literature at the Library of Congress* (30 Jan. 2014). https://blogs.loc.gov/catbird/2014/01/space-time-and-the-poet-sagan/.
Caute, David. *The Dancer Defects*. Oxford: Oxford University Press, 2003.
Cavanagh, Clare. *Lyric Poetry and Modern Politics: Russia, Poland, and the West*. New Haven, CT: Yale University Press, 2009.

Chaikin, Andrew. "Who Took the Legendary Earthrise Photo From Apollo 8?" *Smithsonian Magazine* (Jan. 2018). http://www.smithsonianmag.com/science-nature/who-took-legendary-earthrise-photo-apollo-8-180967505/.
Chakrabarty, Dipesh. "Postcolonial Studies and the Challenge of Climate Change." *New Literary History* 43:1 (Winter 2012): 1–18.
Chakrabarty, Dipesh. "Climate and Capital: On Conjoined Histories." *Critical Inquiry* 41:1 (2014): 1–23.
Chakrabarty, Dipesh. "The Planet: An Emergent Humanist Category." *Critical Inquiry* 46:1 (Autumn 2019): 1–31.
Chakraborty, Sumita. "The Trouble You Promised: Reading Tracy K. Smith." *Los Angeles Review of Books* (26 Aug. 2018). https://lareviewofbooks.org/article/the-trouble-you-promised-reading-tracy-k-smith/.
Chakraborty, Sumita. "Of New Calligraphy: Seamus Heaney, Planetarity, and Lyric's Uncanny Space-Walk." *Cultural Critique* 14:104 (Summer 2019): 101–34.
Chakraborty, Sumita. *Arrow*. Farmington, ME: Alice James Books, 2020.
Chasar, Mike. *Poetry Unbound: Poems and New Media from the Magic Lantern to Instagram*. New York: Columbia University Press, 2020.
Cheah, Pheng. *What Is a World? On Postcolonial Literature as World Literature*. Durham, NC: Duke University Press, 2016.
Chu, Seo-Young. *Do Metaphors Dream of Literal Sleep?* Cambridge, MA: Harvard University Press, 2010.
Clarke, Arthur C. "Astronautics and Poetry." In *The Coming of the Space Age: Famous Accounts of Man's Probing of the Universe*. New York: Meredith Press, 1967, 293–7.
Clinton, Bill. "Remarks to the Community in Derry, Northern Ireland." 30 Nov. 1995. https://www.govinfo.gov/content/pkg/PPP-1995-book2/pdf/PPP-1995-book2-doc-pg1809.pdf.
Collins, Michael. *Carrying the Fire*. New York: Farrar, Straus and Giroux, 1974.
Cook, James R. *The Journals of Captain Cook*. Edited by Philip Edwards. New York: Routledge, 2000.
Corcoran, Neil. *Poets of Modern Ireland: Text, Context, Intertext*. Carbondale: University of Southern Illinois Press, 1999.
Cosgrove, Denis. *Apollo's Eye: A Cartographic Genealogy of the Earth in the Western Imagination*. Baltimore, MD: Johns Hopkins University Press, 2001.
Costello, Bonnie. "Elizabeth Bishop's Impersonal Personal." *American Literary History* 15:2 (2003): 334–66.
Costello, Bonnie. *Planets on Tables: Poetry, Still Life, and the Turning World*. Ithaca, NY: Cornell University Press, 2008.
Costello, Bonnie. *The Plural of Us: Poetry and Community in Auden and Others*. Princeton, NJ: Princeton University Press, 2017.
Crawford, Robert, ed. *Contemporary Poetry and Contemporary Science*. Oxford: Oxford University Press, 2006.
Crowe, Michael J. *The Extraterrestrial Life Debate 1750–1900*. Cambridge: Cambridge University Press, 1986.
Culler, Jonathan. *Theory of the Lyric*. Cambridge, MA: Harvard University Press, 2015.
Cummings, E. E. *E. E. Cummings: Selected Poems*. New York: Liveright, 2007.
Danielson, Dennis. *Paradise Lost and the Cosmological Revolution*. Cambridge: Cambridge University Press, 2014.
Daw, Sarah. *Writing Nature in Cold War American Literature*. Edinburgh, UK: Edinburgh University Press, 2018.

deGrasse Tyson, Neil. "'When I Heard the Learn'd Astronomer' with Neil deGrasse Tyson." 14 Nov. 2013. Video. https://www.amnh.org/exhibitions/permanent/hayden-planetarium/dark-universe.
Deloughrey, Elizabeth. "Satellite Planetarity and the Ends of the Earth." *Public Culture* 26:2 (March 2014): 257–80.
Deloughrey, Elizabeth. *Allegories of the Anthropocene*. Durham, NC: Duke University Press, 2019.
Dery, Mark. "Black to the Future: Interviews with Samuel R. Delany, Greg Tate, and Tricia Rose." In *Flame Wars: The Discourse of Cyberculture*. Edited by Mark Dery. Durham, NC: Duke University Press, 1994, 179–222.
Deutsch, Babette. "To the Moon, 1969." *New York Times* (21 July 1969).
De Villiers, Nicholas. *Opacity and the Closet: Queer Tactics in Foucault, Barthes, and Warhol*. Minneapolis: University of Minnesota Press, 2012.
Dick, Steven J. "Why We Explore." *NASA* (21 July 2005). https://www.nasa.gov/exploration/whyweexplore/Why_We_13.html.
Dickey, James. "Apollo 11: As it Happened – Poet James Dickey, 'The Moon Ground.'" *American Broadcasting Company* (1969). Video, 4:10. https://www.youtube.com/watch?v=zaSGs8DQ_PQ.
Dickey, James. "Off to the Moon." *Life, Special Issue* (4 July 1969).
Dickey, James. "So Long." *Life* (10 Jan. 1969).
Dickinson, Emily. *The Complete Poems of Emily Dickinson*. Edited by Thomas H. Johnson. New York: Little, Brown & Co., 1976.
Dimock, Wai Chee. *Through Other Continents: American Literature Across Deep Time*. Princeton, NJ: Princeton University Press, 2006.
Dimock, Wai Chee. "Gilgamesh's Planetary Turns." In *The Planetary Turn: Rationality and Geoaesthetics in the Twenty-First Century*. Edited by Amy J. Elias and Christian Moraru. Evanston, IL: Northwestern University Press, 2015, 125–42.
Donne, John. *John Donne: The Complete English Poems*. Edited by A. J. Smith. London: Penguin, 1996.
Dove, Rita. *Selected Poems*. New York: Vintage, 1993.
Du Bois, W. E. B. *The Souls of Black Folk*. New York: Norton, 1999.
Dukas, Helen, and Banesh Hoffmann, eds. *Albert Einstein, The Human Side: Glimpses from His Archives*. Princeton, NJ: Princeton University Press, 2016.
Dutta, Mary Buhl. "'Very bad poetry, Captain': Shakespeare in Star Trek." *Extrapolation* 36:1 (1995): 38–45.
Eagleton, Terry. *After Theory*. New York: Basic Books, 2003.
Ebury, Katherine. "'In this valley of dying stars': Eliot's Cosmology." *Journal of Modern Literature* 35:3 (Spring 2012): 139–57.
Edelman, Lee. "The Geography of Gender: Elizabeth Bishop's 'In the Waiting Room.'" *Contemporary Literature* 26:2 (1985): 179–96.
Edelman, Lee. *No Future: Queer Theory and the Death Drive*. Durham, NC: Duke University Press, 2004.
Edmond, Jacob. *A Common Strangeness: Contemporary Poetry, Cross-Cultural Encounter, Comparative Literature*. New York: Fordham University Press, 2012.
Edmond, Jacob. *Make It the Same: Poetry in the Age of Global Media*. New York: Columbia University Press, 2019.
Eiseley, Loren. *The Invisible Pyramid*. Lincoln: University of Nebraska Press, 1970.
Elias, Amy J., and Christian Moraru. *The Planetary Turn: Rationality and Geoaesthetics in the Twenty-First Century*. Evanston, IL: Northwestern University Press, 2015.

Eliot, T. S. "Reflections on Vers Libre." *New Statesman* (3 March 1917).
Eliot, T. S. *The Sacred Wood: Essays on Poetry and Criticism*. London: Faber & Faber, 1997.
Eliot, T. S. *The Complete Prose of T.S. Eliot: The Critical Edition*. Edited by Ronald Schuchard. Baltimore, MD: Johns Hopkins University Press, 2021.
Elliott, Charles. "Elizabeth Bishop: Minor Poet with Major Fund of Love." *Life, Special Issue: Off to the Moon* (4 July 1969).
Ellis, Jonathan. "Reading Bishop Reading Darwin." In *John Holmes, Science in Modern Poetry*. Liverpool, UK: Liverpool University Press, 2012, 181–9.
Emerson, Ralph Waldo. "The American Scholar: An Oration, Delivered Before the Phi Beta Kappa Society, at Cambridge, August 31, 1837." In *The Collected Works of Ralph Waldo Emerson*, Volume I. Edited by Robert E. Spiller. Cambridge, MA: Harvard University Press, 1971, 52–70.
Erwin, Lee. *Star Trek: The Original Series*. Season 3, episode 14, "Whom Gods Destroy." Directed by Herb Wallerstein, featuring William Shatner, Leonard Nimoy, and DeForest Kelley. Aired 29 Nov. 1968.
Faggan, Robert. *Czesław Miłosz: Conversations*. Edited by Peggy Whitman Prenshaw. Jackson, MS: University Press of Mississippi, 2006.
Fisher, Marc. "1968: The year America unraveled." *Washington Post* (29 May 2018). https://www.washingtonpost.com/graphics/2018/national/1968-history-major-events-in-pop-culture/#:~:text=1968%20was%20the%20year%20the%20center%20did%20not,yet%20also%20a%20time%20of%20passion%20and%20possibility.
Folsom, Ed. "When I Heard the Learn'd Astronomer." In *Walt Whitman: An Encyclopedia*. Edited by E. R. LeMaster and Donald D. Kummings. New York: Garland Publishing, 1998, 769.
Foucault, Michel. *The History of Sexuality, Volume 1: An Introduction*. New York: Vintage Books, 1978.
Foucault, Michel. *Discipline and Punish: The Birth of the Prison*. Translated by Alan Sheridan. New York: Vintage Books, 1995.
Fountain, Gary, and Peter Brazeau, eds. *Remembering Elizabeth Bishop: An Oral Biography*. Amherst: University Press of Massachusetts, 1994.
Franaszek, Andrzej. *Miłosz: A Biography*. Cambridge, MA: The Belknap Press of Harvard University Press, 2017.
Frost, Robert. *The Poetry of Robert Frost: The Collected Poems*. Edited by Edward Connery Lathem. New York: Henry Holt, 2002.
Fuller, R. Buckminster. *Operating Manual for Spaceship Earth*. Carbondale: Southern Illinois University Press, 1969.
Galchen, Rivka. "The Eighth Continent: The new race to the moon, for science, profit, and pride." *New Yorker* (29 Apr. 2019): 46–53.
Genette, Gérard. "The Architext." In *The Lyric Theory Reader*. Edited by Virginia Jackson and Yopie Prins. Baltimore: Johns Hopkins University Press, 2014, 17–29.
George, Alice. "The Sad, Sad Story of Laika, the Space Dog, and Her One-Way Trip into Orbit." *Smithsonian Magazine* (11 Apr. 2018). https://www.smithsonianmag.com/smithsonian-institution/sad-story-laika-space-dog-and-her-one-way-trip-orbit-1-180968728/.
Ghosh, Amitav. "The Ghat of the Only World: Agha Shahid Ali in Brooklyn." In *Mad Heart Be Brave: Essays on the Poetry of Agha Shahid Ali*, edited by Kazim Ali. Ann Arbor: University of Michigan Press, 2017, 199–214.
Gilroy, Paul. *Small Acts: Thoughts on the Politics of Black Cultures*. London: Serpent's Tail, 1973.
Gilroy, Paul. *Postcolonial Melancholia*. New York: Columbia University Press, 2004.

Ginsberg, Allen. *The Fall of America: Poems of These States*. San Francisco, CA: City Lights, 1974.
Giragosian, Sarah. "Elizabeth Bishop's Evolutionary Poetics." *Interdisciplinary Literary Studies: A Journal of Criticism and Theory* 13:4 (2016): 475–500.
Glissant, Édouard. *Caribbean Discourse: Selected Essays*. Translated by J. Michael Dash, CARAF Books. Charlottesville: University of Virginia Press, 1989.
Glissant, Édouard. *Philosophie de la Relation* or *Poetics of Relation*. Translated by Betsy Wing. Ann Arbor: University of Michigan Press, 1997.
Glissant, Édouard. "Conversation with Édouard Glissant Aboard the Queen Mary II." Conducted by Manthia Diawara. Translated by Christopher Winks. In *Afro-Modern: Journeys Through the Black Atlantic*. Edited by Tanya Barson and Peter Gorschlüter. London: Tate Publishing, 2010, 58–63.
Glissant, Édouard. "The Thinking of the Opacity of the World." *Philosophie de la relation. Poésie en étendue*, 2009. Translated by Franck Loric, reprinted in *Frieze* d/e 7 (Winter 2012): 77.
Goldstein, Amanda Jo. *Sweet Science: Romantic Materialism and the New Logics of Life*. Chicago, IL: Chicago University Press, 2017.
Goldstein, Laurence. "'The End of All Our Exploring': The Moon Landing and Modern Poetry." *Michigan Quarterly Review* 18:2 (1979): 192–216.
Goldstein, Laurence, and Robert Chrisman, eds. *Robert Hayden: Essays on the Poetry*. Ann Arbor: University of Michigan Press, 2013.
Gorman, Amanda. "Our Purpose in Poetry: Or, Earthrise." *Sierra Club: Los Padres Chapter* (3 Feb. 2021). https://www.sierraclub.org/los-padres/blog/2021/02/earthrise-poem-amanda-gorman.
Gray, Richard. *A History of American Poetry*. Malden, MA: Wiley Blackwell, 2015.
Grobe, Christopher. *The Art of Confession: The Performance of Self from Robert Lowell to Reality TV*. New York: New York University Press, 2017.
Guillory, John. *Cultural Capital: The Problem of Literary Canon Formation*. Chicago, IL: University of Chicago Press, 1993.
Halberstam, Jack, published as Judith Halberstam. *In Queer Time and Place: Transgender Bodies, Subcultural Lives*. New York: New York University Press, 2005.
Hall, Donald. "Goatfoot, Milktongue, Twinbird." In *Goatfoot, Milktongue, Twinbird: Interviews, Essays, and Notes on Poetry, 1970–76*. Ann Arbor: University of Michigan Press, 1978, 117–29.
Haraway, Donna J. *Modest_Witness@Second_Milennium.FemaleMan©_Meets_ OncoMouse: Feminism and Technoscience*. New York: Routledge, 1997.
Harjo, Joy. "The Whole of Time." *On Being with Krista Tippet* (13 May 2021). https://onbeing.org/programs/joy-harjo-the-whole-of-time/.
Hashmi, Shadab Zeest. "'Who will inherit the last night of the past?' Agha Shahid Ali's Architecture of Nostalgia as Translation." In *Mad Heart Be Brave: Essays on the Poetry of Agha Shahid Ali*. Edited by Kazim Ali. Ann Arbor: University of Michigan Press, 2017, 183–9.
Haven, Cynthia L. "A Sacred Vision: An Interview with Czesław Miłosz." *Georgia Review* 57:2 (Summer 2003): 303–14.
Haven, Cynthia L., ed. *Czesław Miłosz: Conversations*. Jackson: University Press of Mississippi, 2006.
Hayden, Robert. *Collected Poems*. Edited by Frederick Glaysher. New York: Liveright, 2013.
Heaney, Seamus. "Current Unstated Assumptions." *Critical Inquiry* 7:4 (Summer 1981): 645–51.

Heaney, Seamus. *Station Island*. New York: Farrar, Straus and Giroux, 1986.
Heaney, Seamus. *The Government of the Tongue: Selected Prose 1978-1987*. New York: Farrar, Straus and Giroux, 1988.
Heaney, Seamus. "Light Finally Enters the Black Hole." *Sunday Tribune* (4 Sept. 1994): A9.
Heaney, Seamus. *The Redress of Poetry*. New York: Farrar, Straus and Giroux, 1995.
Heaney, Seamus. *The Spirit Level*. New York: Farrar, Straus and Giroux, 1996.
Heaney, Seamus. *Opened Ground: Selected Poems, 1966-1996*. New York: Farrar, Straus and Giroux, 1998.
Heaney, Seamus. *Electric Light*. London: Faber & Faber, 2001.
Heaney, Seamus. "In Gratitude for All the Gifts." *Guardian* (10 Sept. 2004). https://www.theguardian.com/books/2004/sep/11/featuresreviews.guardianreview25.
Heaney, Seamus. "Making Sense of a Life." Interviewed by Tiago Moura. *NewsHouse* (13 Apr. 2010). https://www.youtube.com/watch?v=s7sskc1pi_k.
Heaney, Seamus. *District and Circle*. London: Faber & Faber, 2011.
Heaney, Seamus. "A soul on the washing line." *Economist* (5 Sept 2013). https://www.economist.com/prospero/2013/09/05/a-soul-on-the-washing-line.
Heaney, Seamus, trans. *Aeneid Book VI*. New York: Farrar, Straus and Giroux, 2016.
Heidegger, Martin. "The Age of the World Picture." In *The Question Concerning Technology and Other Essays*. Translated by William Lovitt. New York: Harper, 1977, 115-54.
Heidegger, Martin. "'Only a God Can Save Us': Der Spiegel's Interview with Martin Heidegger (1966)." In *The Heidegger Controversy: A Critical Reader*. Edited by Richard Wolin. Cambridge, MA: MIT Press, 1992, 105-6.
Heidegger, Martin. "…Poetically Man Dwells…" In *Poetry, Language, Thought*. Translated by Albert Hofstadter. New York: Harper Perennial Modern Thought, 2001, 209-25.
Heinemann, Arthur. *Star Trek: The Original Series*. Season 3, episode 11, "Wink of an Eye." Directed by Jud Taylor, featuring William Shatner, Leonard Nimoy, and DeForest Kelley. Aired November 29, 1968.
Heise, Ursula K. *Sense of Place and Sense of Planet: The Environmental Imagination of the Global*. Oxford: Oxford University Press, 2008.
Hena, Omaar. *Global Anglophone Poetry: Literary Form and Social Critique in Walcott, Muldoon, de Kok, and Nagra*. New York: Palgrave Macmillan, 2015.
Hena, Omaar. "Globalization and Postcolonial Poetry." In *The Cambridge Companion to Postcolonial Poetry*. Edited by Jahan Ramazani. Cambridge: Cambridge University Press, 2017, 249-62.
Henchman, Anna. *The Starry Sky Within: Astronomy and the Reach of the Mind in Victorian Literature*. Oxford: Oxford University Press, 2014.
Henry, Elisabeth. *Orpheus and His Lute: Poetry and the Renewal of Life*. Carbondale and Edwardsville: Southern Illinois University Press, 1992.
Hetherington, Norriss S., ed. *Cosmology: Historical, Literary, Philosophical, Religious, and Scientific Perspectives* (New York: Garland, 1993).
Hoff, D. B. "History of the Teaching of Astronomy in American High Schools." In *Proceedings of IAU Colloq. 105, held in Williamstown, MA, 27-30 July 1988*. Edited by J. M. Pasachoff and J. R. Percy. Cambridge: Cambridge University Press, 1990, 249-53.
Hoff, Syd. "How do you expect the children to respect you if you don't get time off for good behavior?" Cartoon. *New Yorker* (7 July 1951). http://www.sydhoff.org/pages/TheNewYorker.html.
Hofmann, Richie. "Innovation in Conversation (Part IV): Speaking with Randall Mann, Richie Hofmann, Phillip B. Williams, and Chen Chen." Interviewed by Dora Malech.

Kenyon Review (27 Feb. 2017). http://www.kenyonreview.org/2017/02/innovation-conversation-part-iv-speaking-randall-mann-richie-hofmann-phillip-b-williams-chen-chen/.

Hollister, Susannah L. "Elizabeth Bishop's Geographic Feeling." *Twentieth-Century Literature* 58:3 (Fall 2012): 399–438.

Horrocks, R. "Jeremiah Horrocks, astronomer and poet." *Journal of Royal Society of New Zealand* 42:2 (Sept. 2011): 113–20.

Hoskin, Michael. *Discoverers of the Universe: William and Caroline Herschel*. Princeton, NJ: Princeton University Press, 2011.

Hunter, Walt. *Forms of a World: Contemporary Poetry and the Making of Globalization*. New York: Fordham University Press, 2019.

Jackson, Virginia. *Dickinson's Misery: A Theory of Lyric Reading*. Princeton, NJ: Princeton University Press, 2005.

Jackson, Virginia, and Yopie Prins, eds. *The Lyric Theory Reader: A Critical Anthology*. Baltimore, MD: Johns Hopkins University Press, 2014.

Jaggi, Satya Dev. "After the Moon Mission of Apollo 8." In *The Earthrise and Other Poems*. [Delhi:] Falcon Poetry Society, 1969.

Jaggi, Satya Dev. *The earthrise and other poems*. [Delhi:] Falcon Poetry Society, 1969.

Jaggi, Satya Dev. *One looks earthward again*. [Delhi:] Falcon Poetry Society, 1970.

Jaggi, Satya Dev. *Our awkward earth*. [Delhi:] Falcon Poetry Society, 1970.

Jaggi, Satya Dev. *The moon voyagers and other poems*. [Delhi:] Falcon Poetry Society, 1970.

Jameson, Fredric. *Archaeologies of the Future: The Desire Called Utopia and Other Science Fictions*. London: Verso, 2005.

Javadizadeh, Kamran. "The Atlantic Ocean Breaking on Our Heads: Claudia Rankine, Robert Lowell, and the Whiteness of the Lyric Subject." *PMLA* 134:3 (May 2019): 475–90.

Jazeel, Tariq. "Spatializing Difference Beyond Cosmopolitanism: Rethinking Planetary Futures." *Theory, Culture, & Society* 28:5 (Sept. 2011): 75–97.

Johnson, Barbara. "Apostrophe, Animation, Abortion." *Diacritics* 16:1 (Spring 1986): 28–47.

Johnson, L. B. "Evaluation of Space Program" (28 Apr. 1961). In *Exploring the Unknown: Selected Documents in the History of the U.S. Civil Space Program*. Edited by John M. Logsdon. Washington, D.C.: NASA, 2004.

Kafer, Alison. "Compulsory Bodies: Reflections on Heterosexuality and Able-bodiedness." *Journal of Women's History* 15:3 (2003): 77–89.

Kant, Immanuel. *Critique of Judgment*. Translated by Werner Pluhar. Indianapolis, IN: Hackett, 1987.

Kay, Magdalena. *In Gratitude for All the Gifts: Seamus Heaney and Eastern Europe*. Toronto, ON: University of Toronto Press, 2012.

Keats, John. *John Keats: The Major Works*. Edited by Elizabeth Cook. Oxford: Oxford University Press, 1990.

Kelly, Scott. *Endurance: My Year in Space, A Lifetime of Discovery*. New York: Knopf, 2017.

Kennard, Emily. "What's In A Name?", Glenn Research Center, 15 May 2009. https://www.nasa.gov/centers/glenn/about/history/silverstein_feature.html.

Kennedy, John F. "If the Soviets Control Space, They Can Control Earth." *Missiles and Rockets* 7:15 (10 Oct. 1960): 12–13.

Kennedy, John F. "Address at Rice University on the Nation's Space Effort." Speech, Rice University, Houston, Texas (12 Sept. 1962). Johnson Space Center. https://er.jsc.nasa.gov/seh/ricetalk.htm.

Kessler, Elizabeth A. *Picturing the Cosmos: Hubble Space Telescope Images and the Astronomical Sublime*. Minneapolis: University of Minnesota Press, 2012.

Kestenbaum, David. "So Over the Moon." In "Episode 655: The Not-So-Great Unknown." Hosted by Ira Glass. Produced by WBEZ. *This American Life* (24 Aug. 2018). Podcast, MP3 audio. https://www.thisamericanlife.org/655/the-not-so-great-unknown.

King, Bruce. "Agha Shahid Ali's Tricultural Nostalgia." *Journal of South Asian Literature* 29:2 (1994): 1–20.

Kirsch, Adam. "Czesław Miłosz's Battle for Truth." *New Yorker* (22 May 2017). http://www.newyorker.com/magazine/2017/05/29/Czesław-Miłoszs-battle-for-truth.

Knickerbocker, Scott. *Ecopoetics: The Language of Nature, the Nature of Language*. Amherst: University of Massachusetts Press, 2012.

Knox, Helene. "Space Poems: Close Encounters Between the Lyric Imagination and 25 Years of NASA Space Exploration." *Lunar Bases and Space Activities of the 21st Century*. Edited by W. W. Mendell. Houston, TX: The Lunar and Planetary Institute, 1985.

Komunyakaa, Yusef. *Pleasure Dome: New and Collected Poems*. Middletown, CT: Wesleyan University Press, 2001.

Komunyakaa, Yusef. "Journey into '[American Journal]'." In *Robert Hayden: Essays on the Poetry*. Edited by Laurence Goldstein and Robert Chrisman. Ann Arbor: University of Michigan Press, 2013, 332–4.

Kubrick, Stanley, director. *2001: A Space Odyssey*. Metro-Goldwyn-Mayer, 1968.

Lai, Jennifer. "We're Bringing 1,100 Haiku to Mars This Fall." *Slate* (9 Aug. 2013). https://slate.com/technology/2013/08/maven-haiku-nasa-s-spacecraft-will-be-bringing-1100-poems-to-mars-this-november.html.

Lal, P., ed. *Modern Indian Poetry in English: An Anthology & a Credo*, 2nd ed. Kolkata: Writers Workshop, 1971.

Latour, Bruno. *An Inquiry into the Mode of Existence: An Anthropology of the Moderns*. Paris: Éditions La Découverte, 2012.

Latour, Bruno. "Anti-Zoom." In *Scale in Literature and Culture (Geocriticism and Spatial Literary Studies)*. Edited by Michael T. Clarke and David Wittenberg. Cham, Switzerland: Palgrave Macmillan, 2017, 93–101.

Lazier, Benjamin. "Earthrise; or, The Globalization of the World Picture." *American Historical Review* 116:3 (2011): 602–30.

Leithauser, Emily. "Remembering Poetry: Figures of Scale in the Postwar Anglophone Lyric." Doctoral dissertation, Emory University, 2016.

Leithauser, Emily. "Traveling Figures and Figures of Travel in 'The Arkansas Testament': Derek Walcott's Quarrel with the American South." *The Global South* 10:1 (Spring 2016): 85–106.

León, Christina. "Forms of Opacity: Roaches, Blood, and Being Stuck in Xandra Ibarra's Corpus." *ASAP/Journal* 2:2 (May 2017): 369–92.

Leonard, Philip. "Message to the gods: the space poetry that transcends human rivalries." *The Conversation* (15 Nov. 2017). https://theconversation.com/message-to-the-gods-the-space-poetry-that-transcends-human-rivalries-86572.

Leonard, Philip. *Orbital Poetics: Literature, Theory, World*. London: Bloomsbury Academic, 2019.

Lewis, S. L., and M. A. Maslin. "Defining the Anthropocene." *Nature* 519 (Mar. 2015): 171–80.

Locke, John. *Locke: Two Treatises of Government*. Edited by Peter Laslett. Cambridge: Cambridge University Press, 1960.

Loff, Sarah. "Haikus Selected To Be Carried On NASA's Next Mission to Mars." *NASA* (8 Aug. 2013). https://www.nasa.gov/content/haikus-selected-to-be-carried-on-nasas-next-mission-to-mars/.

Loff, Sarah. "Poem by American Matriarch Flown on Orion Presented to NASA Administrator." *NASA* (6 Apr. 2015). https://www.nasa.gov/content/poem-by-american-matriarch-flown-on-orion-presented-to-nasa-administrator.

Longenbach, James. *Modern Poetry After Modernism*. Oxford: Oxford University Press, 1997.

Longinus. *On the Sublime*. Translated by H. L. Havell. London: Macmillan, 1890.

Longley, Edna. "'Inner Emigré' or 'Artful Voyeur'? Seamus Heaney's *North*." *Poetry in the Wars*. Newcastle upon Tyne, UK: Bloodaxe Books, 1986.

Longley, Michael. *Collected Poems*. London: Jonathan Cape, 2006.

Lowe, Lisa. *The Intimacies of Four Continents*. Durham, NC: Duke University Press, 2015.

Lowell, Robert. "Robert Lowell, The Art of Poetry No. 3." Interviewed by Frederick Seidel. *Paris Review* 25 (Winter–Spring 1961). https://www.theparisreview.org/interviews/4664/the-art-of-poetry-no-3-robert-lowell.

Lowell, Robert. *History*. New York: Farrar, Straus and Giroux, 1973.

Luna, Joe. "Space / Poetry." *Critical Inquiry* 43:1 (Autumn 2016): 110–38.

MacArthur, Marit. "One World? The Poetics of Passenger Flight and the Perception of the Global." *PMLA* 127:2 (Mar. 2012): 264–82.

MacLeish, Archibald. "Riders on the Earth." *New York Times* (25 Dec. 1968). https://timesmachine.nytimes.com/timesmachine/1968/12/25/issue.html.

MacLeish, Archibald. "Voyage to the Moon." *New York Times* (21 July 1969). https://archive.nytimes.com/www.nytimes.com/library/national/science/nasa/072169sci-nasa.html.

Maher, Neil M. *Apollo in the Age of Aquarius*. Cambridge, MA: Harvard University Press, 2017.

Marché, Jordan D. *Theaters of Time and Space: American Planetaria, 1930–1970*. Rutgers, NJ: Rutgers University Press, 2005.

Materer, Timothy. "Mirrored Lives: Elizabeth Bishop and James Merrill." *Twentieth Century Literature* 51:2 (Summer 2005): 179–209.

May, Elaine Tyler. *Homeward Bound: American Families in the Cold War Era*. New York: Basic Books, 1988.

McCabe, Susan. "Survival of the Queerly Fit: Darwin, Marianne Moore, and Elizabeth Bishop." *Twentieth Century Literature* 55: 4, Darwin and Literary Studies (Winter 2009), 547–71.

McCarthy, Mary. "Symposium: I Would Like to Have Written…." In *Elizabeth Bishop and Her Art*. Edited by Lloyd Schwartz and Sybil B. Estess. Ann Arbor: University of Michigan Press, 1983, 267–9.

McDougal, Walter A. …*the Heavens and the Earth: A Political History of the Space Age*. New York: Basic Books, 1985.

McGuirk, Kevin. "A. R. Ammons and the Whole Earth." *Cultural Critique* 37 (Autumn 1997): 131–58.

McIntosh, Hugh. "Conventions of Closeness: Realism and the Creative Friendship of Elizabeth Bishop and Robert Lowell." *PMLA* 112:2 (Mar. 2012): 231–47.

McLemee, Scott. "Sunspots and Poetry." *Inside Higher Ed* (7 June 2019). https://www.insidehighered.com/views/2019/06/07/review-tracy-daugherty-dante-and-early-astronomer-science-adventure-and-victorian.

McMorris, Neville. *The Nature of Science*. Vancouver, BC: Fairleigh Dickinson University Press, 1989.

Medovoi, Leerom. "Cold War American Culture as the Age of Three Worlds." *Minnesota Review* 55–7 (2002): 167–86.

Mendelson, Edward. "'So Huge a Phallic Triumph': Why Apollo Had Little Appeal for Auden." *New York Review of Books* (12 Aug. 2019). https://www.nybooks.com/daily/2019/08/12/so-huge-a-phallic-triumph-why-apollo-had-little-appeal-for-auden/.

Meredith, George. "Lucifer in Starlight." In Laura Emma Lockwood, *Sonnets, Selected from English and American Authors*. Boston, MA: Houghton Mifflin Co., 1916.

Merrill, James. "Elizabeth Bishop, 1911–1979." In *Elizabeth Bishop and Her Art*. Edited by Lloyd Schwartz and Sybil P. Estess. Ann Arbor: University of Michigan Press, 1983, 259–62.

Merrill, James. *A Different Person: A Memoir*. San Francisco, CA: Harper, 1994.

Merrill, James. *Collected Poems*. New York: Knopf, 2002.

Merrill, James. *Selected Poems*. Edited by J. D. McClatchy and Stephen Yenser. New York: Knopf, 2008.

Messeri, Lisa. "Beyond the Anthropocene: Un-Earthing an Epoch." *Environment and Society: Advances in Research* 6 (2015): 28–47.

Messeri, Lisa. *Placing Outer Space: An Earthly Ethnography of Other Worlds*. Durham, NC: Duke University Press, 2016.

Middleton, Peter. *Physics Envy: American Poetry and Science in the Cold War and After*. Chicago, IL: The University of Chicago Press, 2015.

Miéville, China. "Cognition as Ideology: A Dialectic of SF Theory." In *Red Planets: Marxism and Science Fiction*. Edited by Mark Bould and China Miéville. Middletown, CT: Wesleyan University Press, 2009, 231–48.

Mill, John Stuart. *Essays on Poetry*. Edited by F. Parvin Sharpless. Columbia: University of South Carolina Press, 1976.

Millier, Brett. *Elizabeth Bishop: Life and the Memory of It*. Berkeley: University of California Press, 1993.

Miłosz, Czesław. *Native Realm: A Search for Self-Definition*. Translated by Catherine S. Leach. New York: Farrar, Straus and Giroux, 1968.

Miłosz, Czesław. *The Year of the Hunter*. Translated by Madeline G. Levine. New York: Farrar, Straus and Giroux, 1994.

Miłosz, Czesław. *New and Collected Poems 1931–2001*. New York: Ecco, 2001.

Miłosz, Czesław. *Second Space*. Translated by Robert Hass. New York: HarperCollins, 2005.

Miłosz, Czesław. *The Mountains of Parnassus: A Novel*. Translated by Stanley Bill. New Haven, CT: Yale University Press, 2017.

Milton, John. *The Complete Poems*. New York: Penguin, 1999.

Mlinko, Ange. "The Lyric Project: On Tracy K. Smith and Cathy Park Hong." *Parnassus: Poetry in Review* 33: 1–2 (2013): 417–26, 432.

Moretti, Franco. *Distant Reading*. New York: Verso, 2013.

Morgan, Peter. "Moondust." *The Crown*, Season 3, Episode 7. Directed by Jessica Hobbs, featuring Olivia Colman, Tobias Menzies, and Helena Bonham Carter. Aired 17 Nov. 2019, https://www.netflix.com/watch/80215737?trackId=200257859.

Morton, Timothy. *The Ecological Thought*. Cambridge, MA: Harvard University Press, 2010.

Morton, Timothy. "Sublime Objects." *Speculations II*. Santa Barbara, CA: punctum books, 2011, 207–27.

Morton, Timothy. *Hyperobjects: Philosophy and Ecology after the End of the World*. Minneapolis: University of Minnesota Press, 2013.

Moskowitz, Clara. "Astronomer's Poem Mourns Ailing Kepler Spacecraft." *In Brief: Space.com* (16 May 2013). https://www.space.com/21179-kepler-telescope-failure-poem.html.

Mullen, Harryette, and Stephen Yenser. "Theme and Variations of Robert Hayden's Poetry." In *Robert Hayden: Essays on the Poetry*. Edited by Laurence Goldstein and Robert Chrisman. Ann Arbor: University of Michigan Press, 2013, 233–50.
Nabokov, Vladimir. *Pale Fire*. New York: Vintage, 1962.
Nadel, Alan. *Containment Culture: American Narratives, Postmodernism, and the Atomic Age*. Durham, NC: Duke University Press, 1995.
NASA. "Astronomers Confounded by Massive Rocky World," 2 June 2014, *NASA*, https://www.nasa.gov/ames/kepler/astronomers-confounded-by-massive-rocky-world.
Needham, Lawrence. Interview with Agha Shahid Ali. *The Verse Book of Interviews: 27 Poets on Language, Craft & Culture*. Edited by Brian Henry and Andrew Zawacki. Amherst, MA: Verse Press, 2005.
Nelson, Deborah. *Pursuing Privacy in Cold War America*. New York: Columbia University Press, 2002.
North, John. *Norton History of Astronomy and Cosmology*. New York: Norton, 1995.
O'Donoghue, Bernard. "Heaney and the Public." *The Cambridge Companion to Seamus Heaney*. Cambridge: Cambridge University Press, 2009.
O'Driscoll, Dennis. *Stepping Stones: Interviews with Seamus Heaney*. New York: Farrar, Straus and Giroux, 2010.
Oliver, Kelly. "The Earth's Refusal: Heidegger." In *Earth and World: Philosophy After the Apollo Missions*. New York: Columbia University Press, 2015, 111–62.
Orr, M. A. *Dante and the Early Astronomers*. London: Gall and Inglis, 1913.
Page, Michael. "The Darwin Before Darwin: Erasmus Darwin, Visionary Science, and Romantic Poetry." *Papers on Language and Literature* 41:2 (2005): 146–69.
Parrinder, Patrick, ed. *Learning from Other Worlds: Estrangement, Cognition, and the Politics of Science Fiction and Utopia*. Durham, NC: Duke University Press, 2001.
Patterson, William H. *Robert A. Heinlein: In Dialogue with His Century, Volume 2: The Man Who Learned Better, 1948–1988*. New York: Tor, 2016.
PBS. "Apollo's Daring Mission," *NOVA*. 1968.
Perloff, Majorie. *Poetic License: Essays on Modernist and Postmodernist Lyric*. Evanston, IL: Northwestern University Press, 1990.
Phillis, Jen Hedler. *Poems of the American Empire: Lyric Form in the Long Twentieth Century*. Iowa City: University of Iowa Press, 2019.
Pickard, Zachariah. *Elizabeth Bishop's Poetics of Description*. Montreal, QC: McGill-Queen's University Press, 2009.
Plath, Sylvia. *The Collected Poems*. New York: HarperCollins, 2008.
Poole, Robert. *Earthrise: How Man First Saw Earth*. New Haven: Yale University Press, 2009.
Popova, Maria. "Planeterium: Astrophysicist Janna Levin Reads Adrienne Rich's Tribute to Trailblazing Women in Science." *BrainPickings* (27 Apr. 2017). https://www.brainpickings.org/2017/04/27/janna-levin-reads-planetarium-by-adrienne-rich/.
Popova, Maria. "A Brave and Startling Truth: Astrophysicist Janna Levin Reads Maya Angelou's Stunning Humanist Poem That Flew to Space, Inspired by Carl Sagan." *BrainPickings* (9 May 2018). https://www.brainpickings.org/2018/05/09/a-brave-and-startling-truth-maya-angelou/.
Posmentier, Sonya. *Cultivation and Catastrophe: The Lyric Ecology of Modern Black Literature*. Baltimore, MD: Johns Hopkins University Press, 2017.
Pound, Ezra. "What I Feel About Walt Whitman." In *Whitman: A Collection of Critical Essays*. Edited by Roy Harvey Pearce. Englewood Cliffs, NJ: Prentice Hall, 1962, 8.

Pound, Ezra. *Selected Prose, 1909–1965*. Edited by William Cookson. London: Faber & Faber, 1973.
Pound, Ezra. *ABC of Reading*. London: Faber & Faber, 1991.
President's Science Advisory Committee. "Introduction to Outer Space." White House Document. 26 Mar. 1958.
Price, Ruth. *The Lives of Agnes Smedley*. Oxford: Oxford University Press, 2005.
Quashie, Kevin. *The Sovereignty of Quiet: Beyond Resistance in Black Culture*. New Brunswick, NJ: Rutgers University Press, 2012.
Quinn, Justin. *Between Two Fires: Transnationalism and Cold War Poetry*. Oxford: Oxford University Press, 2015.
Ramazani, Jahan. *Poetry of Mourning: The Modern Elegy from Hardy to Heaney*. Chicago, IL: University of Chicago Press, 1994.
Ramazani, Jahan. *A Transnational Poetics*. Chicago, IL: University of Chicago Press, 2009.
Ramazani, Jahan. *Poetry and Its Others: News, Prayer, Song, and the Dialogue of Genres*. Chicago, IL: University of Chicago Press, 2013.
Ramazani, Jahan. "Seamus Heaney's Globe." *The Irish Review* 49/50 (Winter–Spring 2014/15): 38–53.
Ranft, Erin. "The Afrofuturist Poetry of Tracie Morris and Tracy K. Smith." *Journal of Ethnic American Literature* 4 (2014): 75–85.
Rankine, Claudia. "Introduction" to *The Collected Poems of Adrienne Rich*. New York: Norton, 2016, xxxvii–xlvii.
Rauscher, Judith. "On Common Ground: Translocal Attachments and Transethnic Affiliations in Agha Shahid Ali's and Arthur Sze's Poetry of the American Southwest." *European Journal of American Studies* 9:3 (2014), n.p.
Rein, Lisa, and Michael Horowitz. "Inner Space and Outer Space: Carl Sagan's Letters to Timothy Leary (1974)." Timothy Leary Archives. http://www.timothylearyarchives.org/carl-sagans-letters-to-timothy-leary-1974/.
Rich, Adrienne. "When We Dead Awaken: Writing as Re-Vision." *College English* 34:1 (Oct. 1972): 18–32.
Rich, Adrienne. "Compulsory Heterosexuality and Lesbian Existence." *Signs* 5:4 (Summer 1980): 631–60.
Rich, Adrienne. "Eye of the Outsider: The Poetry of Elizabeth Bishop." *Boston Review* 8.2 (1983): 15–17.
Rich, Adrienne. *What Is Found There: Notebooks on Poetry and Politics*. New York: Norton, 2003.
Rich, Adrienne. *The Collected Poems of Adrienne Rich*. New York: Norton, 2016.
Rich, Adrienne. *Essential Essays: Culture, Politics, and the Art of Poetry*. Edited by Sandra M. Gilbert. New York: Norton, 2018.
Richardson, Mattie Udora. "No More Secrets, No More Lies: African American History and Compulsory Heterosexuality." *Journal of Women's History* 15:3 (2003): 63–76.
Riordan, Maurice, and Jocelyn B. Burnell. *Dark Matter: Poems of Space*. London: Calouste Gulbenkian Foundation, 2008.
Roman, Camille. *Elizabeth Bishop's World War II–Cold War View*. New York: Palgrave Macmillan, 2001.
Ronda, Margaret. *Remainders: American Poetry at Nature's End*. Stanford, CA: Stanford University Press, 2018.
Rosaldo, Renato. "Imperialist Nostalgia." *Representations* 26 (1989): 107–22.
Rosenthal, A. M. "Standby Update Moon Poem." *New York Times* (18 July 1989). https://www.nytimes.com/1989/07/18/opinion/on-my-mind-standby-update-moon-poem.html.

Rubin, Andrew N. *Archives of Authority: Empire, Culture, and the Cold War*. Princeton, NJ: Princeton University Press, 2012.
Ruefle, Mary. *Madness, Rack, and Honey: Collected Lectures*. Seattle, WA, and New York: Wave Books, 2012.
Rusert, Britt. *Fugitive Science: Empiricism and Freedom in Early African American Culture*. New York: New York University Press, 2017.
Sachs, Wolfgang. *Planet Dialectics: Explorations in Environment and Development*. London: Zed Books, 1999.
Sagan, Carl. *Cosmos*. New York: Ballantine, 1980.
Sappho. *Sappho*. Translated by Mary Barnard. Berkeley: University of California Press, 2012.
Schneiderman, Jason. "The Loved One Always Leaves: The Poetic Friendship of James Merrill and Agha Shahid Ali." *American Poetry Review* 43:5 (September 2014): 11–12.
Schuster, Joshua. "Another Poetry Is Possible: Will Alexander, Planetary Futures, and Exopoetics." *Resilience: A Journal of the Environmental Humanities* 4:2–3 (Spring–Fall 2017): 147–65.
Scott-Heron, Gil. "Whitey on the Moon." Recorded 1970. Track 9 on *Small Talk at 125th and Lenox*. New York: Flying Dutchman Records, vinyl LP.
Shalev, Eran. "'A Republic Amidst the Stars': Political Astronomy and the Intellectual Origins of the Stars and Stripes." *Journal of the Early Republic* 31 (Spring 2011): 39–73.
Shelley, Percy Bysshe. *A Defense of Poetry*. Boston: Ginn & Co., 1840.
Shiban, John, writer. *Breaking Bad*. Season 3, episode 6, "Sunset." Directed by John Shiban, featuring Bryan Cranston, Aaron Paul, and David Costabile. Aired 25 Apr. 2010.
Shklovsky, Viktor. "Art as Technique." In *Literary Theory: An Anthology*. Edited by Julie Rivkin and Michael Ryan. Hoboken, NJ: Blackwell, 2004, 15–21.
Shoaib, Mahwash. "'The Grief of Broken Flesh: The Dialectic of Desire and Death in Agha Shahid Ali's Lyrics." In *Mad Heart Be Brave: Essays on the Poetry of Agha Shahid Ali*. Edited by Kazim Ali. Ann Arbor: University of Michigan Press, 2017, 170–82.
Shockley, Evie. *Renegade Poetics: Black Aesthetics and Formal Innovation in African American Poetry*. Iowa City: University of Iowa Press, 2011.
Siddiqi, Asif A. "Competing Technologies, National(ist) Narratives, and Universal Claims: Towards a Global History of Space Exploration." *Technology and Culture* 51:2 (2010): 425–43.
Silliman, Ron. "Wednesday, July 7, 2010," Silliman's Blog. http://www.ronsilliman.blogspot.com/2010/07/i-know-whenever-i-use-phrase-school-of.html.
Smith, J. Y. "Christine Laitin Dies at 65." *Washington Post* (6 Apr. 1995). https://www.washingtonpost.com/archive/local/1995/04/06/christine-laitin-dies-at-65/62c0b636-aee2-479d-81a5-5ea52179e5cc/.
Smith, Tracy K. "'Something We Need': An Interview with Tracy K. Smith." Interviewed by Charles H. Rowell. *Callaloo* 27:4 (Fall 2004): 858–72.
Smith, Tracy K. *Life on Mars*. Minneapolis, MN: Graywolf, 2011.
Smith, Tracy K. "'Life on Mars' Author Explores Humans' Relationship with Universe Through Poetry." Interviewed by Gwen Ifill. *PBS Newshour* (16 May 2011). http://www.pbs.org/newshour/bb/entertainment-jan-june11-tracysmith_05-16/.
Smith, Tracy K. "Imagining the Universe: Life on Mars with Tracy K. Smith." Talk at Stanford University (1 Dec. 2014). http://www.youtube.com/watch?v=QW7WIJaSfl4&t=110s
Smith, Tracy K. *Ordinary Light*. New York: Vintage, 2015.
Smith, Tracy K. "Moving toward What I Don't Know': An Interview with Tracy K. Smith." Interviewed by Claire Schwartz. *Iowa Review* 26:2 (Fall 2016). https://iowareview.org/

from-the-issue/volume-46-issue-2-—-fall-2016/moving-toward-what-i-dont-know-interview-tracy-k-smith.

Smith, Tracy K. "Meet our new U.S. Poet Laureate, Tracy K. Smith." Interviewed by Carolyn Kellogg. *Los Angeles Times* (13 June 2017). www.latimes.com/books/jacketcopy/la-et-jc-poet-laureate-20170614-htmlstory.html.

Smith, Tracy K. "Political Poetry Is Hot Again. The Poet Laureate Explores Why, and How." *New York Times* (10 Dec. 2018). http://www.nytimes.com/2018/12/10/books/review/political-poetry.html.

Smith, Tracy K., and David DeVorkin. "Clockwork Universe: Emily Dickinson Birthday Tribute." Lecture at the Folger Shakespeare Library, Washington, D.C., 12 Dec. 2016.

Snow, C. P. *The Two Cultures and the Scientific Revolution*. Cambridge: Cambridge University Press, 1959.

Spaide, Christopher. "'A Delicate, Vibrating Range of Difference': Adrienne Rich and the Postwar Lyric 'We.'" *College Literature* 47:1 (2020): 89–124.

Spigel, Lynn. *Welcome to the Dream House: Popular Media and the Postwar Suburbs*. Durham, NC: Duke University Press, 2001.

Spivak, Gayatri. *Death of a Discipline*. New York: Columbia University Press, 2003.

Spivak, Gayatri. *An Aesthetic Education in the Era of Globalization*. Cambridge, MA: Harvard University Press, 2012.

Steinman, Lisa. *Made in America: Science, Technology, and American Modernist Poets*. New Haven: Yale University Press, 1987.

Stening, Tanner. "This Covering of Blue." *Rattle* (17 Oct. 2021). https://www.rattle.com/this-covering-of-blue-by-tanner-stening/.

Stevens, Wallace. *The Palm at the End of the Mind: Selected Poems and a Play*. Edited by Holly Stevens. New York: Vintage, 1967.

Stilling, Robert. *Beginning at the End: Decadence, Modernism, and Postcolonial Poetry*. Cambridge, MA: Harvard University Press, 2018.

Stone, Robert, writer. *American Experience*. Season 31, episode 4, "Chasing the Moon, Part Two." Directed by Robert Stone, featuring Buzz Aldrin, George Alexander, and Bill Anders. Aired 10 July 2019. https://www.pbs.org/video/chasing-the-moon-part-2-7s7mhp/.

Su, John S. *Ethics and Nostalgia in the Contemporary Novel*. Cambridge: Cambridge University Press, 2005.

Suhr-Sytsma, Nathan. *Poetry, Print, and the Making of Postcolonial Literature*. Cambridge: Cambridge University Press, 2017.

Sullivan, Hannah. "Still Doing It by Hand: Auden and the Typewriter." In *Auden at Work*. Edited by Bonnie Costello and Rachel Galvin. New York: Palgrave Macmillan, 2015, 19–23.

Sullivan, Robert. *Captain Cooke in the Underworld*. Auckland, NZ: Auckland University Press, 2003.

Suvin, Darko. *Defined by a Hollow: Essays on Utopia, Science Fiction and Political Epistemology*. Bern, Switzerland: Peter Lang, 2010.

Suvin, Darko. *The Metamorphoses of Science Fiction: On the Politics and History of a Literary Genre*. Bern, Switzerland: Peter Lang, 2016.

Swanner, Leandra Altha. "Mountains of Controversy: Narrative and the Making of Contested Landscapes in Postwar American Astronomy." Doctoral dissertation, Harvard University, 2013. https://dash.harvard.edu/bitstream/handle/1/11156816/Swanner_gsas.harvard_0084L_10781.pdf?sequence=3.

Tageldin, Shaden M. "Reversing the Sentence of Impossible Nostalgia: The Poetics of Postcolonial Migration in Sakinna Boukhedenna and Agha Shahid Ali. *Comparative Literature Studies* 40:2 (2003): 232–64.

Tally, Robert T., Jr. "Beyond the Flaming Walls of the World: Fantasy, Alterity, and the Postcolonial Constellation." In *The Planetary Turn: Rationality and Geoaesthetics in the Twenty-First Century*. Edited by Amy J. Elias and Christian Moraru. Evanston, IL: Northwestern University Press, 2015, 193–210.

Tennyson. *The Complete Poetical Works of Tennyson*. Edited by W. J. Rolfe. Boston: Houghton Mifflin Co., 1989.

Teveleva, Irina. "James Dickey and the Apollo Program." Washington University in St. Louis University Libraries. 18 July 2019. https://library.wustl.edu/james-dickey-and-the-apollo-program/.

Thoreau, Henry David. *The Writings of Henry David Thoreau, Volume 5*. Edited by Bradford Torrey. Boston, MA: Houghton, Mifflin & Co., 1851–2.

Tracy, Robert. "Westering: Seamus Heaney's Berkeley Year." *California Magazine* (10 Jan. 2018). https://alumni.berkeley.edu/california-magazine/online/westering-seamus-heaneys-berkeley-year/.

Travisano, Thomas. *Midcentury Quartet: Bishop, Lowell, Jarrell, Berryman, and the Making of a Postmodern Aesthetic*. Charlottesville, VA, and London: University Press of Virginia, 1999.

Trethewey, Natasha. *Native Guard*. Boston, MA: Mariner Books, 2007.

Tribbe, Matthew D. *No Requiem for the Space Age: The Apollo Moon Landings and American Culture*. Oxford: Oxford University Press, 2014.

Trousdale, Rachel. *Nabokov, Rushdie, and the Transnational Imagination: Novels of Exile and Alternate Worlds*. New York: Palgrave Macmillan, 2010.

Vas Dias, Robert. *Inside Outer Space: New Poems of the Space Age*. New York: Anchor Books, 1970.

Vendler, Helen. *Invisible Listeners: Lyric Intimacy in Herbert, Whitman, and Ashbery*. Princeton, NJ: Princeton University Press, 2005.

Walcott, Derek. *The Poetry of Derek Walcott, 1948–2013*. New York: Farrar, Straus and Giroux, 2014.

Walls, Laura Dassow. *Passage to Cosmos: Alexander von Humboldt and the Shaping of America*. Chicago, IL: University of Chicago Press, 2009.

Walls, Laura Dassow. "O Pioneer." *American Scientist* 104:2 (March–April 2016): 118.

Ward, David. "The Space of Poetry: Inhabiting Form in the Ghazal." *University of Toronto Quarterly* 82:1 (2013): 62–71.

Warner, Michael. *Publics and Counterpublics*. New York: Zone Books, 2010.

Weaver, Kenneth F. "Historic Color Portrait of Earth from Space." *National Geographic* 132:5 (1967).

Wells, Helen T., Susan H. Whiteley, and Carrie E. Karegeannes. *Origins of NASA Names*. The NASA History Series. Washington, D.C.: National Aeronautics and Space Administration, 1976. https://history.nasa.gov/SP-4402.pdf.

Welna, David. "Space Force Bible Blessing at National Cathedral Sparks Outrage." (13 Jan. 2020). https://www.npr.org/2020/01/13/796028336/space-force-bible-blessing-at-national-cathedral-sparks-outrage.

Wheatley, Phillis. *Phillis Wheatley: Complete Writings*. Edited by Vincent Carretta. New York: Penguin, 2001.

White, Frank. *The Overview Effect: Space Exploration and Human Evolution*. Reston, VA: American Institute of Aeronautics and Astronautics, 2014.

White, Gillian. "Words in Air and 'Space' in Art: Bishop's Midcentury Critique of the United States." In *Elizabeth Bishop in the 21st Century: Reading the New Editions*, edited by Angus J. Cleghorn, Bethany Hicok, and Thomas J. Travisano. Charlottesville: University of Virginia Press, 2012, 255–73.

White, Gillian. *Lyric Shame: The "Lyric" Subject of Contemporary American Poetry.* Cambridge, MA: Harvard University Press, 2014.
Whiting, Melanie. "50 Years Ago: Considered Changes to Apollo 8." *NASA History* (8 Aug. 2018). https://www.nasa.gov/feature/50-years-ago-considered-changes-to-apollo-8.
Whitman, Walt. *The Complete Poems.* Edited by Francis Murphy. New York: Penguin, 2004.
Wientzen, Timothy. "Not a Globe but a Planet: Modernism and the Epoch of Modernity." *Modernism/Modernity* 2:4 (2018). https://doi.org/10.26597/mod.0039.
Wilford, John N. "Astronauts Land on Plain; Collect Rocks; Plant Flag." *New York Times* (21 July 1969). https://archive.nytimes.com/www.nytimes.com/library/national/science/nasa/072169sci-nasa.html.
Wright, Pearce, et al. "The Colour of Space: Earth Photographed by Man from over the Moon." *The Times* (6 Jan. 1969).
Young, Kevin. *Bunk: The Rise of Hoaxes, Humbug, Plagiarists, Phonies, Post-Facts, and Fake News.* Minneapolis, MN: Graywolf, 2017.
Zimmerman, Brett. "Nineteenth-Century American Astronomy and the Sublime." *Journal of the Royal Astronomical Society of Canada* 97:3 (June 2003), 120–3.

Index

For the benefit of digital users, indexed terms that span two pages (e.g., 52–53) may, on occasion, appear on only one of those pages.

Adorno, Theodor 66
Ali, Agha Shahid 2, 17–18, 34, 48–9, 100, 103, 109–33, 136
  "The Country Without a Post Office" 115–16
  "From Amherst to Kashmir" 117–18, 132–3
  "For You" 131
  "Lunarscape" 109–13, 115–16, 128–9, 131–3, 136, 199
  "Ghazal" 117
  "The Keeper of the Dead Hotel" 126
  "Postcard from Kashmir" 114–15
  "Snow on the Desert" 129–31
American exceptionalism 2, 11–12, 19, 34–5, 173–4
American imperialism 2–3, 5–7, 34–5, 39–40, 42, 77–8, 136, 157–8; and American identity 19, 82–3
American nationalism 1–2, 3n.6, 5–6, 76–7
Angelou, Maya 200–1
  "A Brave and Startling Truth" 200
Anglo–American empire 2–3, 6–7, 34, 102–3
Anthropocene 12–13, 33–4, 36–7, 41, 44–7, 105–6, 123–4
Apollo 8 24–5, 28, 34, 36, 51–2, 58–9, 63–6
apostrophe 6–10, 12–13, 32–4, 45–51, 54–5, 59, 65–6, 75–6, 97–8, 100, 133, 136, 138–9, 145–6, 164–5, 197
Armstrong, Neil 5, 101–2, 104–5
Asghar, Fatima, "Pluto Shits on the Universe" 32–3
astronauts 24n.78, 28, 33–4, 36–9, 47–9, 51–9, 63, 65–6, 69–72, 82, 101–4, 106–8, 131–2, 135–6, 144, 148–50, 157, 159, 187
astronomical poem 20–2, 30–2, 47–8, 68, 79–80

astronomy 1–2, 5–7, 10–16, 19–26, 30–2, 42, 68, 79, 82–3, 86–100, 123–4, 128–9, 137–8, 146–8, 150–1, 156, 159, 163–4, 167–8, 171–5, 177–80, 182, 185–6, 189–90
Auden, W.H. 13–15, 49–51, 55–6, 75, 80–1, 107–8, 131–2, 143–4, 202
  "Funeral Blues" 202
  "In Memory of W.B. Yeats" 146–8
  "The More Loving One" 13–14
  "Ode to Terminus" 14–15

Bakhtin, Mikhail 8–9, 146–8
Baraka, Amiri, "Planetary Exchange" 173–4
Barbauld, Anna 16–17
Bishop, Elizabeth 2, 33–4, 67–8, 73–9, 87–97, 99–100, 179–80, 195–6, 199
  "The Armadillo" 88, 92
  "Insomnia" 88, 91–2
  "In the Waiting Room" 49–50, 68, 94–100
  "The Shampoo" 88, 91–2
Blue Marble 36–8, 40–1, 45–6, 53–6, 64, 194, 200–1
Brodsky, Joseph 139–44, 146–8, 156–7
  "The Condition We Call Exile" 146
  "On Grief and Reason" 146
  "A Poet and Prose" 146–8
  "Ninety Years Later" 146–8
  To Urania 146
Brooks, Cleanth 8–9
Burke, Edmund 15–16, 167–8
Burnell, Jocelyn Bell 84–7
Byrd, Jodi 2–3, 128–30

capitalism 36–7, 43, 69–70, 77–8, 127, 148–9, 194
Central Intelligence Agency (CIA) 70–1, 122–3

Chakrabarty, Dipesh 43–5
Chakraborty, Sumita 32–3, 47–8, 160, 164–5
   "Marigolds" 32–3
civil rights movement 27–30, 102–3, 108–9, 181–2
Cold War 1–2, 5–6, 8–9, 11–12, 22–34, 40–1, 45–6, 48–9, 56–7, 68, 70–9, 88, 96, 100, 103, 105–7, 109–12, 118–33, 135–6, 138–48, 152–3, 188
Cold War America 5, 7–8, 69–70, 90
confessional poetry 68–75, 77–81, 120–1, 152–3, 175–6
Cosgrove, Denis 52–4, 56–7, 95–6
cosmic harmony 3–5, 19–21, 32, 60–1, 129–30, 172
   see also cosmos
cosmic scale 13–17, 34–5, 93
   see also deep time
cosmopolitanism 41, 114, 117, 139–40, 142–8, 160–2
cosmos 3–5, 8–11, 13–16, 18, 20–2, 30–2, 34, 46, 53–4, 56, 60–3, 82–3, 86, 99–100, 106–7, 112–15, 127, 131–2, 134–5, 138–48, 160–2, 165–9, 171–3, 194–5, 199, 201–2
   see also cosmic harmony
cosmology 4–5, 8–9, 13–15, 22, 36–7, 53–4, 60–1, 105–6, 128–30, 136, 146–8, 164
   heliocentric cosmos, and 4–5
   see also Ptolemy
containment culture 33–4, 68–71, 91, 120–1
Culler, Jonathan 7–8, 12–13, 46–7, 59, 97–8
cultural imperialism 42–3, 123

Dante, *The Divine Comedy* 4–5, 30–2, 199–200
dark matter 185–6
Darwin, Charles 18, 88–90, 98–9
deep time 42, 65–6, 91–2, 123–4, 197–8
deGrasse Tyson, Neil 30–2
Dickey, James 27–8, 52–3, 55–6, 58–60, 65–6, 199
Dickinson, Emily 124–5, 177–9
Donne, John 9–10
Dove, Rita 169–70, 178–80, 183–6
   "Geometry" 183–4

Earthrise 28, 36–7, 40–1, 45–9, 51–2, 55–66, 79–80, 104–7, 109–11, 123–4, 132–3, 200–1
Earth System Science 43
Einstein, Albert 22, 34–5, 134–5, 138–9, 142–3, 184–5
elegy 7–8, 34, 39, 49–50, 59, 67, 75, 103, 115, 118–20, 123, 130–3, 146–8, 166–9, 180–5, 189–202
Eliot, T.S. 22, 25–6, 30–2, 70–1, 101, 107–8, 115, 122–3, 138–9, 141–4, 183
   Four Quartets 25–6, 101–2, 138–9, 143
   "Tradition and the Individual Talent" 22
Emerson, Ralph Waldo 19
Endeavour 2–4, 128–9
environmentalism 36–7
Eurocentrism 2–3, 170–1, 176–7

Faiz, Ahmad Faiz 115, 118–20, 123–6
   "We Who Are Executed" 118–20
formal verse 70–1, 76–7, 120–4
Foucault, Michel 74–5
free verse 20–1, 68–71, 76–7, 91–2, 120–4, 131–2

genre 5–12, 34, 37–9, 47–8, 76–7, 86–7, 103, 146, 148–50, 154–5, 169–70, 179–80, 186, 190–2
ghazal 7–8, 109–11, 116–33
Gilroy, Paul 46–7, 102–3, 173–4
Ginsberg, Allen 48
Glissant, Édouard 170–1
globalization 12–13, 40–5, 56, 109–10, 112, 114, 117–18, 129–30
global culture 12–13
Global South 122–3
gravity 32–4, 139–48, 155–7, 159, 161–4
   see also relativity

Harjo, Joy 200–1
Hayden, Robert 30–2, 173–6, 181–2
   "[American Journal]" 173–5
Heaney, Seamus 2, 17–18, 115, 135–44, 146–8, 152–66, 169–70, 179–83
   "Alphabets" 47–50, 136–7, 164–6, 199
   "The Birthplace" 143–4
   "*Clearances* VIII" 180–1
   "The Cure at Troy" 154–5
   "Digging" 137–8, 159–60

INDEX   223

"The Flight Path" 162-3
"Kinship" 156-7
"Out of This World" 166
"The Tollund Man" 160-1
"Whatever You Say, Say Nothing" 161-2
"Westering" 157-9, 161-2
Heidegger, Martin 43-4, 50
Horrocks, Jeremiah, *Venus in sole visa* 3-4
Hubble Space Telescope 7-8, 167-8, 171, 189-90
Humboldt, Alexander von 18-19, 21, 30-2

imperialism 2-3, 5-7, 33-5, 39-40, 42-4, 48, 56, 77-8, 102-3, 120-1, 123-9, 136, 141-2, 145-6, 157-8, 185-6
imperialist nostalgia 102-3, 127
India 48-9, 56, 102-3, 109-13, 118-23, 126-30, 133, 144-5
individualism 17-18, 69-72, 82, 159-60, 175-6
Ireland 115-16, 137, 157, 161-2, 182

Jackson, Virginia 40
Jaggi, Satya Dev 48-50, 109-10
Johnson, Barbara 12-13, 46-7

Kant, Immanuel 15-16, 167-8
Kashmir 110-18, 120-1, 127, 129-30, 132-3, 136
Keats, John 8-11, 29-32, 101-2
  "On First Looking into Chapman's Homer" 62-3
Kennedy, John F. 23-4, 39, 69-71, 106-7, 110-11, 123

Longinus 15-16, 45
Longley, Michael 156
lyric reading 6-7, 13-14, 33-4, 40, 49-50, 60, 65-6, 96, 164-5
lyric subjectivity 22, 52-3, 68, 77-8, 170-1, 175-6, 178-9

MacLeish, Archibald 27-8, 55-8, 60, 79-80, 99, 103-6, 109-10, 199
  "Voyage to the Moon" 58-9, 103-6, 109-10
Mars 36, 80, 92, 189, 200-1
MAVEN (Mars Atmosphere and Volatile EvolutioN) craft 1

Meredith, George, "Lucifer in Starlight" 60-2, 98-9
Merrill, James 67, 75-6, 91-2, 124-5, 199
  "Overdue Pilgrimage to Nova Scotia" 67, 75
Miłosz, Czesław 115, 134-43, 146-54, 156-7, 165-6
  "Ars Poetica?" 149-50
  "Campo Dei Fiori" 137-8, 165-6
  "Incantation" 151-5
  The Mountains of Parnassus 135, 148-51
Milton, John 4-5, 20-1, 29-32, 61, 146-8
  Paradise Lost 4-5, 20-1, 61, 146-8
moon 18, 21-5, 27-8, 32-4, 36, 43-8, 51-3, 55-9, 63, 68-71, 79-80, 88, 91-4, 100-12, 115-20, 123-6, 128-33, 136, 157-9, 162, 193, 201-2
moon landing 5-6, 12, 24-5, 27-8, 48, 51-2, 55-8, 88-90, 101-2, 107-10, 131-2, 135, 156-9, 161
musica universalis 3-4

NASA 1-3, 5, 23-34, 36-8, 51-9, 64-6, 79-80, 98-9, 101-4, 107-8, 125-6, 187, 199-202
National Defense Education Act 23-4, 82, 95-6
National Geographic 33-4, 37-9, 60-2, 94-9
National Science Foundation 128
New Criticism 7-8, 22, 71, 117, 176-7
Northern Ireland 136-7, 139-42, 152-9, 162-3, 180-1
nostalgia 46, 80-1, 101-3, 106-20, 123-5, 127, 130-2, 157, 193

opacity 34, 61-2, 68, 72-8, 85-9, 94, 100, 168-71, 175-98
  Glissant, Édouard, and 170-1
  queer 73-6, 87-8, 94, 100

Pakistan 120-1, 123, 126
planetarity 41-5
Plath, Sylvia 72-3, 77-9
  "A Birthday Present" 73
  "Wuthering Heights" 72-3
postcolonialism 11-12, 48-9, 100, 102-3, 109-10, 112-13, 120-3, 138, 145-6, 170-1

Pound, Ezra 1, 17–18, 22, 70–1, 98–9, 122–3, 152–3, 169–70, 176–7
Ptolemy 4–5, 8–9, 146–8, 199

Quashie, Kevin 178–9, 183

Ramazani, Jahan 12, 43–5, 47–8, 60, 65–6, 99–100, 109–10, 143–4, 164–5, 191–2
relativity 134–5, 143
  see also Einstein, Albert
Rich, Adrienne 2, 33–4, 45–6, 68, 74–88, 95–6, 124–5, 161, 171
  "The Explorers" 68, 79–81
  "Orion" 81–2
  "Planetarium" 68, 79–82, 84, 86–7, 95–6, 161
  "Snapshots of a Daughter-in-Law" 81
  "When We Dead Awaken" 80–2
Roddenberry, Gene 5, 29–30, 201–2
Romantic poetry 5, 7–8, 10–11, 15–17, 20–1, 24–7, 29–32, 107–8, 118–21, 141–2
Russia 141–2, 161

Sagan, Carl 7–8, 30–2, 199
Sappho 15, 45, 107–10
science fiction 7–8, 29–30, 48, 62–3, 135, 138–9, 148–52, 154–5, 171–2, 175–6, 185–6, 188–98
Scott-Heron, Gil 108–9
Shakespeare, William 8, 29–30
Shelley, Percy Bysshe 25, 29–30, 62–3
Smith, Benedict 1
Smith, Tracy K. 2, 7–8, 13–15, 47–8, 167–70, 175–83, 185–6
  "Life on Mars" 185–6
  "The Museum of Obsolescence" 194
  "My God, It's Full of Stars" 47–8, 168–70, 180, 196
  "Sci-Fi" 192–3, 195–6
  "Solstice" 195–6
  "They May Love All That He Has Chosen and Hate All That He Has Rejected" 193–4
  "The Universe Is a House Party" 188–9
Soviet Union (USSR) 23–4, 48–50, 70–1, 123, 126, 146

space age, the 23–5, 32–4, 39, 48, 50, 68–72, 88–9, 95–6, 99, 145–6, 173
space culture 2–3, 7–8, 30–2
Space Force 1–2, 32–3
space photography 2
  see also Hubble Space Telescope, Blue Marble, Earthrise
space race 1–2, 21–4, 29–30, 39, 43, 51–2, 70–1, 95–6, 106–7, 112–13, 151
Spivak, Gayatri 42–4, 47–8, 99–100, 160, 164–5
Star Trek 2–3, 5, 29–32, 156, 189, 201–2
Stevens, Wallace, "The Planet on the Table" 39
St. Lucia 140–2, 145–6
sublime 15–17, 19, 45, 54–5, 98–9, 149, 167–8, 171–2, 187
Sullivan, Robert, Captain Cook in the Underworld 3–5
surveillance 68, 72–6, 81, 88, 90

Tennyson, Alfred Lord 25–6, 30–2
Tohono O'odham Nation 127–9
Trethewey, Natasha 169–70, 178–80, 183–6
  "Theories of Time and Space" 184–5

Venus 2–4, 30–2, 92, 128–9

Walcott, Derek 2, 139–46
  "Forest of Europe" 140–1
  "The Fortunate Traveller" 49–50, 145–6
  "North and South" 145–6
  "The Schooner Flight" 145–6
Wheatley, Phillis, "On Imagination" 171–2
Whitman, Walt 5, 17–22, 25, 30–2, 70–1, 83–4, 173
  "Kosmos" 18
  "Song of Myself" 17–18
  "When I Heard the Learn'd Astronomer" 20–1, 30–2

Yeats, W.B. 115–16, 136–7, 146–8, 161–2, 165–6, 181–2
Young, Kevin 108–9